An Atlas
of
Mammalian
Viruses

Authors

Erskine L. Palmer, Ph.D.
Supervisory Research Microbiologist
Special Pathogens Branch
Centers for Disease Control
Atlanta, Georgia

Mary Lane Martin
Research Microbiologist
Special Pathogens Branch
Centers for Disease Control
Atlanta, Georgia

CRC Press, Inc.
Boca Raton, Florida

Library of Congress Cataloging in Publication Data

Palmer, Erskine L.
  An atlas of mammalian viruses.

  Bibliography: p.
  Includes index.
  1. Viruses--Atlases. 2. Ultrastructure
(Biology) I. Martin, Mary Lane. II. Title.
[DNLM: 1. Microscopy, Electron--Atlases.
2. Virology--Atlases. QW 17 P173e]
QR363.P34   616'.0194   81-12214
ISBN 0-8493-6628-3   AACR2

This book represents information obtained from authentic and highly regarded sources. Reprinted material is quoted with permission, and sources are indicated. A wide variety of references are listed. Every reasonable effort has been made to give reliable data and information, but the author and the publisher cannot assume responsibility for the validity of all materials or for the consequences of their use.

Direct all inquiries to CRC Press, Inc., 2000 N.W. 24th Street, Boca Raton, Florida, 33431.

International Standard Book Number 0-8493-6628-3

Library of Congress Card Number 81-12214
Printed in the United States

PREFACE

Colleagues and visitors to our laboratory have suggested many times that we prepare an atlas of virology for use as a diagnostic reference source. This book is in response to those requests. We designed the Atlas primarily for investigators in the field of virology interested in the electron microscopic diagnosis of viruses related to human disease. However, we have tried to go beyond the level of a manual and have presented ultrastructural information on some viruses which may serve as a reference guide for those interested in virus morphology and attendant molecular analyses.

We wish to acknowledge the following contributions to the Atlas. For micrographs, we thank Mr. E. H. Cook, Jr., Hepatitis Laboratories, CDC, Phoenix, for Plates 1 and 2 (Hepatitis Viruses); Dr. Joseph Esposito, Virology Division, CDC, Atlanta, for Plate 1 (Picornaviridae); Dr. G. William Gary, Jr., Virology Division, CDC, Atlanta, for Plate 1 (Adenoviridae), Plate 2 (Parvoviridae), and Plates 5 and 6 (miscellaneous, Bacteriophage); Ms. Kathy Hancock, Virology Division, CDC, for Plate 3 (Parvoviridae); Mrs. Alyne Harrison, Virology Division, CDC, for Plate 2 (Adenoviridae); Dr. Fredrick Murphy, College of Veterinary Medicine and Biomedical Sciences, Colorado State University, for Plate 1 (Rhabdoviridae), Plate 1 (Togaviridae), and Plate 7 (Paramyxoviridae); Dr. Russell Regnery, Virology Division, CDC, for Plate 4 (Rhabdoviridae); and Ms. Sylvia Whitfield, Virology Division, CDC, for Plate 6 (Arenaviridae). We also thank those members of the Virology Division, CDC, who provided viral preparations used to make the electron micrographs. We express appreciation to Dr. Walter Dowdle, Assistant Director for Sciences, CDC, Director, Center for Infectious Diseases, CDC, and to Dr. Roy Chamberlain, Director, Virology Division, CDC, for reading the manuscript and making many helpful comments, and to Ms. Patricia Echols for invaluable secretarial assistance.

Erskine L. Palmer
Mary Lane Martin

THE AUTHORS

Erskine Palmer received his B.S. and M.S. degrees from Florida State University and the Ph.D. degree from the University of Mississippi. He has worked as a Research Virologist at the Centers for Disease Control since 1964 and is currently in charge of the Virology Division's Electron Microscopy Laboratory. He has conducted research on a wide variety of viruses of public health importance.

Mary Lane Martin holds a B.S. degree in Biology from Stetson University and a M.S. degree in Biology from Georgia State University. She has worked as a Virologist at the Centers for Disease Control since 1966 and has carried out numerous studies on the morphology of mammalian viruses. She currently works with Dr. Palmer in the Special Pathogens Branch of the CDC.

TABLE OF CONTENTS

Chapter 1

INTRODUCTION

Over the past decade the transmission electron microscope (EM) has become an increasingly valuable tool in diagnostic virology, and new applications of EM are regularly described in scientific literature. This surge in diagnostic usage has roughly followed the development and improvement of negative staining techniques, which have greatly simplified otherwise tedious methods of specimen preparation. In addition, many laborious and time-consuming purification procedures aimed at ridding the specimen of contaminating material before it is put into the microscope column are no longer necessary. Now no body material or fluid is considered too unclean for direct examination by EM; consequently, important advances have been made in the identification of agents of disease that had escaped more conventional methods of detection.

Besides the obvious advantages of simplicity and speed, diagnostic EM with negative staining has the added attribute of producing results which, when positive, are unequivocal. For example, when crusts or vesicle fluids from persons with suspected smallpox are examined, poxvirus can be unequivocally distinguished from herpesvirus. This was one of the earliest and most important applications of diagnostic EM, and it greatly facilitated the surveillance of suspected smallpox as this disease was being eradicated from the world.

Along the same lines, rotavirus in fecal specimens is easily and quickly identified by EM and has been documented as the causative agent of many outbreaks of infantile diarrhea. Other viruses from the gut, especially small viruses in the 20- to 30-nm range, have also been implicated in gastroenteritis and have been identified by the technique of immune EM. The agent of hepatitis A, the Norwalk agent, and several other small virus particles have been visualized by this procedure.

Cutting embedded tissues into thin sections and looking for evidence of virus replication in cells is another method of diagnostic EM, but this procedure is hardly practical for routine diagnosis because it is tedious and lengthy. Therefore, most of the micrographs in this volume have been selected because they illustrate the appearance of the virions when negatively stained and because they show virus structure with which the electron microscopist must be familiar for diagnostic applications. In addition to micrographs delineating the salient features of viruses representative of the various virus groups, the atlas contains many high-resolution micrographs. The text also includes pertinent supplementary information concerning each virus and available data on its fine ultrastructure. Most of the micrographs depict viruses which have been negatively stained, some among a milieu of cellular debris, in order to show the viruses as they appear in clinical material.

Chapter 2

VIRUS CLASSIFICATION

Much progress in virus classification is now being made by the International Committee on Taxonomy of Viruses (ICTV), an agency of the International Association of Microbiological Societies. Study Groups of the ICTV now give careful consideration to the largest possible number of viral characters before recommending taxonomic status.

Viruses of vertebrates can be broadly grouped as having either DNA or RNA genomes. The DNA viruses are further broadly grouped into those which have cubic capsid symmetry and a naked nucleocapsid (families Parvoviridae, Papovaviridae and Adenoviridae) and those which have cubic or unresolved capsid symmetry but are enveloped (families Herpesviridae, Poxviridae). Similarly, the RNA viruses can be primarily grouped into those which have cubic capsid symmetry and a naked nucleocapsid (Picornaviridae, Reoviridae), and those with helical symmetry (Orthomyxoviridae, Paramyxoviridae, Retroviridae, Rhabdoviridae). Other groups of RNA viruses include those which are unsymmetric or with unknown symmetry (Coronaviridae), and those which have cubic capsid symmetry and a complex envelope (Togaviridae). Two groups (Arenaviridae and Bunyaviridae) have ribonucleoprotein nucleocapsids which are circular and segmented. It has not yet been determined whether these are helical structures. The current taxonomic status of viruses is presented by Fenner,[2] and an update of taxonomy from the Fourth International Congress of Virology, The Hague, 1978, has recently been published.[4]

Stylized drawings of most of the viruses presented in this book are shown in Plate 1 to demonstrate relative shape, size, and basic characteristics of each group.

REFERENCES

1. **Fenner, F.,** Classification and nomenclature of viruses. Second report of the ICTV, *Intervirology*, 7, 1, 1976.
2. **Mathews, R. E. R.,** The classification and nomenclature of viruses. Summary of results of meetings of The International Committee on Taxonomy of Viruses in The Hague, Sept. 1978, *Intervirology*, 11, 133, 1979.
3. **Melnick, J.,** Taxonomy of viruses, *Prog. Med. Virol.*, 24, 207, 1978.
4. **Murphy, F. A.,** Taxonomy of vertebrate viruses, in *Handbook of Microbiology*, Laskin, A. I. and Lechevalier, H., Eds., CRC Press, Inc., Boca Raton, Fla., 1978, 623.

DNA VIRUSES

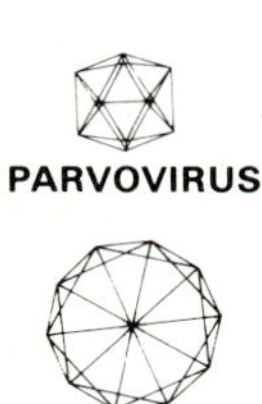

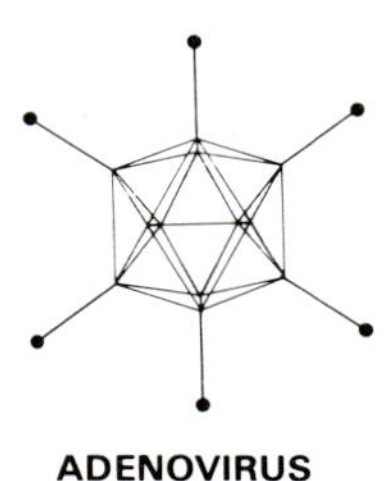

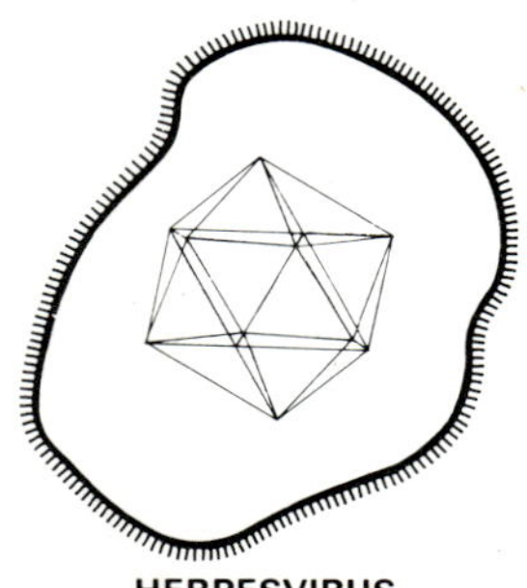

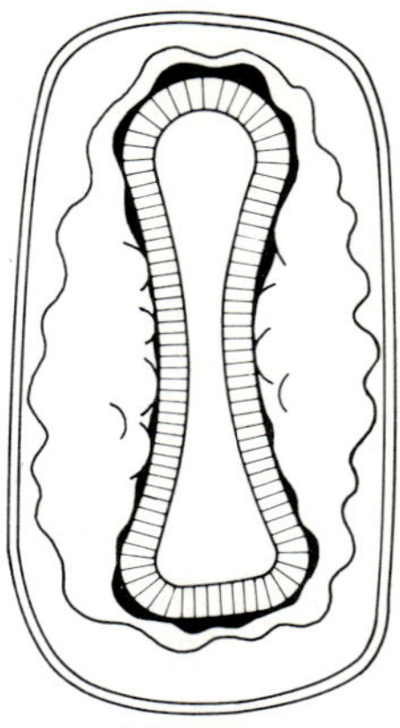

RNA VIRUSES

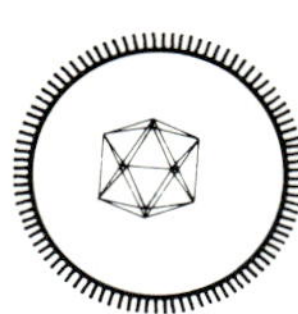

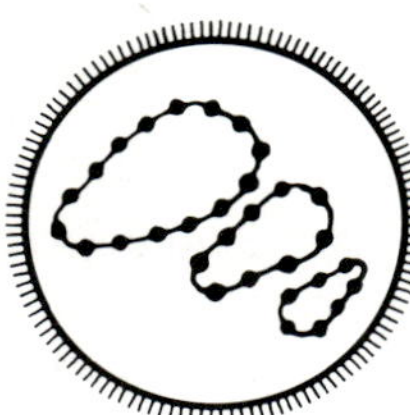

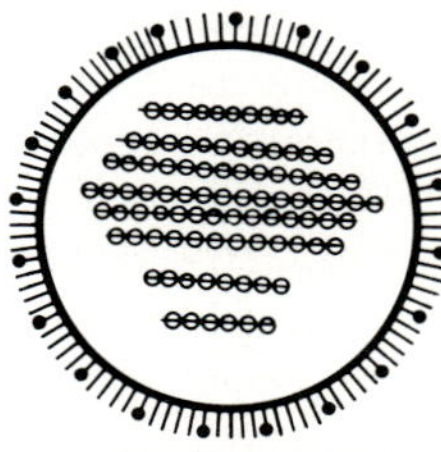

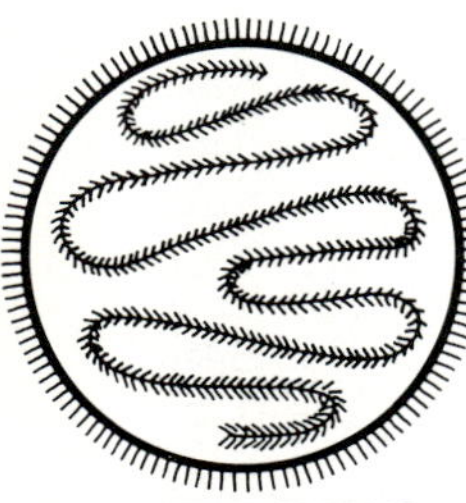

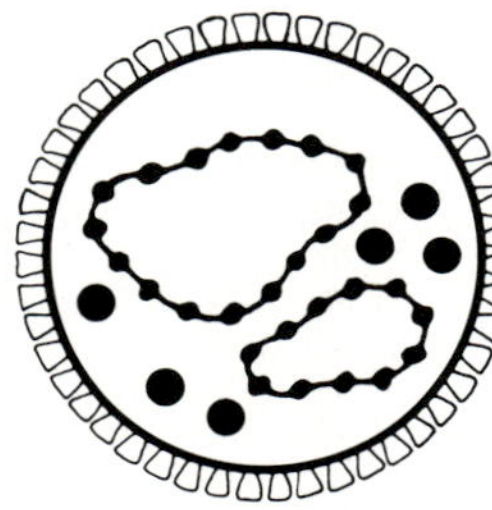

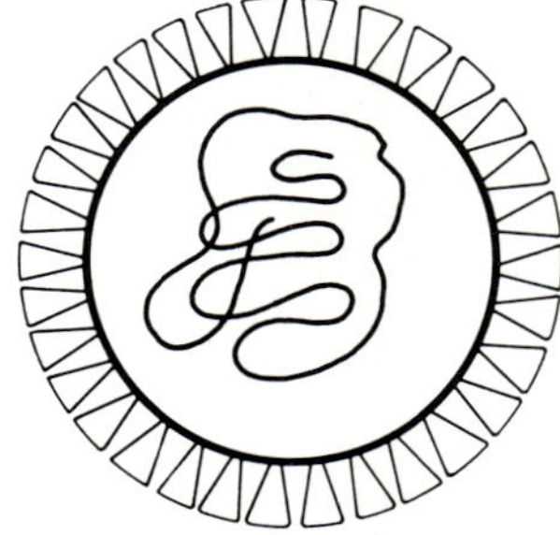

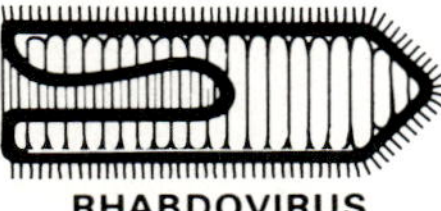

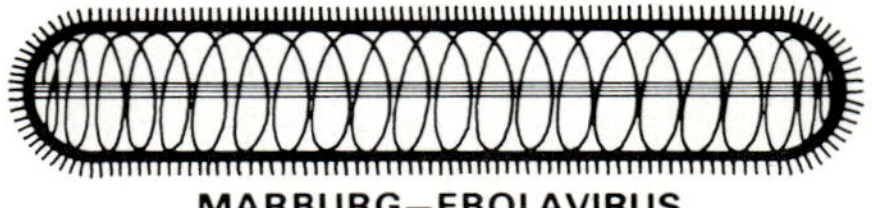

Chapter 3

PRINCIPLES OF STRUCTURE

Rapid improvement of the technology of EM in the late 1950s and early 1960s served to confirm virus structural organization which had been suggested by light microscopy, nucleic acid-protein translation studies, and X-ray analyses. Not only could EM easily discern that some viruses were isometric and others were rod-shaped, but it could also reveal that they were composed of subunits in an orderly arrangement.

In the early 1960s Caspar and Klug[1] proposed the principles that explained the precise regularity of virus structure. First, they postulated that the structure of rod-shaped plant viruses, such as tobacco mosaic virus, was the result of helical symmetry in the arrangement of the protein subunits. It was subsequently found that helical symmetry accounted for the structure of the flexible rods of nucleoprotein contained within the pleomorphic envelopes of certain animal viruses, such as the orthomyxoviruses. Second, they proposed that the isometric viruses were constructed with icosahedral symmetry. With the exception of the poxvirus group, which has a complex brick-shaped structure, all animal viruses thus far discovered fit into one of these two groups.

Icosahedral Symmetry

Icosahedral symmetry, also known as 5:3:2 symmetry because it possesses central axes of fivefold, threefold, and twofold symmetry, results when structures are composed of 60, or certain multiples of 60, subunits. These subunits may be thought of as being arranged on the surface of an actual icosahedron, a geometric figure composed of 20 equilateral triangles and having 12 vertices. In fact, some viruses, such as adenoviruses, are structured in exactly this manner. Others, like reoviruses, are structured in such a way that, although they possess icosahedral symmetry, the base is not a regular icosahedron. These two basic arrangements are designated, respectively, as classes $P = 1$ and $P = 3$ (Figure 1). A third group, with a skew arrangement, is constructed as a $P = 7$. In a skew arrangement, each of the fivefold vertices is set at an angle in respect to the neighboring fivefold vertices. Thus, in some viruses of this class, the vertices may be skewed to the right, whereas in others the vertices may be skewed to the left.

Caspar and Klug[1] showed that clustering of the 60 subunits of isometric viruses into groups, or capsomeres, was plausible and would account for the hexagonal appearance of many viruses and for the number of the capsomeres that had been calculated for some viruses. Clustering into groups of five subunits (pentameres) and six units (hexameres) has been found in all isometric animal viruses thus far delineated by high resolution EM with the possible exception of the Reoviridae. The Reoviridae have a capsid formed by sharing of subunits which results in a prominent pattern of holes. The simplest viruses are composed of 60 subunits comprising 12 pentameres which are located at the vertices of the icosahedron. Larger isometric viruses possess 12 pentameres and also hexameres, whose numbers are always multiples of 10, arranged in a regular array between the pentameres.

Only certain arrangements of pentameres and hexameres satisfy the requirement of icosahedral symmetry. These are designated by their triangulation number, or T-number, which is defined as the number of equilateral subtriangles into which each of the 20 sides of the icosahedron can be divided, each subtriangle containing three protein subunits. The T-number for members of the $P = 1$ class, obtained by counting the number (n) of hexameres between two neighboring pentameres, is calculated by the equation $(n + 1)^2 = T$ (Figure 2). T may therefore be 1, 4, 9, 16, 25, etc. For the P

= 3 and P = 7 classes, hexameres are not aligned in rows between the pentameres, but the T-number may be calculated by the equation $T = Pf^2$, where f is any integer. For the P = 3 class, T may therefore be 3, 12, 27, 48, etc. This equation is also valid for the P = 1 class.

Knowledge of the T-number allows quick calculation of the capsomere number when the structural subunits are clustered into pentameres and hexameres. The number (N) of capsomeres is given by the equation $N = 10T + 2$ or $N = 10(T - 1)$ hexameres + 12 pentameres. Possible T numbers for the P = 1, P = 3, and P = 7 classes would therefore result in capsomere numbers of 12, 32, 42, 72, 92, 122 and so on; the largest number so far discovered for viruses infecting humans is 252 for adenovirus, which possesses T = 25 icosahedral structure.

The T-number also has a direct relationship to the number of structural subunits, which may be expressed as 60T. For example, a virus with 32 capsomeres (P = 3, T = 3) is composed of 180 subunits; 60 of them make up the 12 pentameres, and the remaining 120 comprise the 20 hexameres.

In some viruses, such as reovirus and calicivirus, the pattern of holes is as prominent a morphological feature as are the capsomeres. The symmetrical arrangement of holes is given the triangulation number t, as compared to the triangulation number T for capsomere arrangement.

The capsomeres form a closed shell or capsid, surrounding the nucleic acid core of the virus. The capsid and the enclosed genome are referred to as the nucleocapsid, and the same term is used to describe the helical arrangement of protein and nucleic acid helices in viruses with rod-shaped components. In addition to the basic capsid structure, some icosahedral viruses also possess an envelope, usually a pleomorphic lipid-containing covering studded with small projections. The complete infectious particle is termed the virion. In this atlas, descriptions of viruses include identifying structural details of the viral envelopes and, if known, the number and arrangement of the capsomeres of each isometric virus.

Helical Symmetry

Helical symmetry is the result of identical polypeptide structural units assembled by virtue of their shape so that the bonding from unit to unit produces a curled structure (Figure 3). Secondary bonding of the units comprising each turn of the "curve" to those of the previous turn is an offset arrangement which forms a helix as the nucleic acid is incorporated. The nucleic acid occupies a channel in the center of the protein helix. This helical arrangement of the protein and nucleic acid is called a nucleocapsid. Helical nucleocapsids have a single axis of symmetry which passes longitudinally down the center of the helix. The nucleocapsid of different viruses varies in the periodicity of the helical turns, so that some are relatively rigid and others very flexible. The nucleocapsid of helical animal viruses is further coiled to fit into pleomorphoric lipid-containing viral envelopes which are usually covered with characteristic surface projections or "spikes."

REFERENCES

1. **Caspar, D. L. D. and Klug, A.,** Physical principles in the construction of regular viruses, *Cold Spring Harbor Symp. Quant. Biol.,* 27, 1, 1962.
2. **Caspar, D. L. D.,** Design principles in virus particle construction, in *Viral and Rickettsial Infections in Man,* Horsfall, F. and Tamm, I., Eds., Lippincott, Philadelphia, 1964, 51.
3. **Horne, R. W.,** *Virus Structure,* Academic Press, New York, 1974.
4. **Mattern, C. F. T.,** *The Molecular Biology of Animal Viruses,* Vol. 1, Nayak, D. P., Ed., Marcel Dekker, New York, 1977, 1.

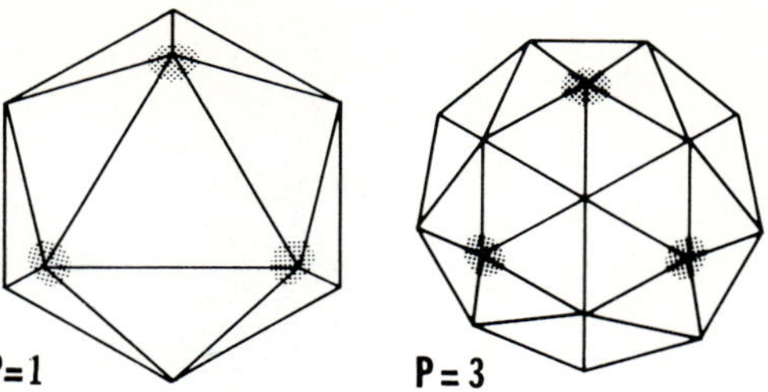

FIGURE 1.   Deltahedra of P = 1 and P = 3 classes of icosahedral symmetry. Differences in the basic structure are evident in the lines connecting adjacent fivefold vertices (shaded). In virions of the P = 1 class, capsomeres are arranged in a straight line between the fivefold vertices. In virions of the P = 3 class, capsomeres are located in a staggered arrangement between the fivefold vertices.

FIGURE 2.   Diagram of the arrangement of capsomeres of the capsid of herpes simplex virus. There are three hexameres between two neighboring pentameres (shaded). The T number calculated by the equation $(n + 1)^2 = T$ is therefore 16. There are 162 capsomeres.

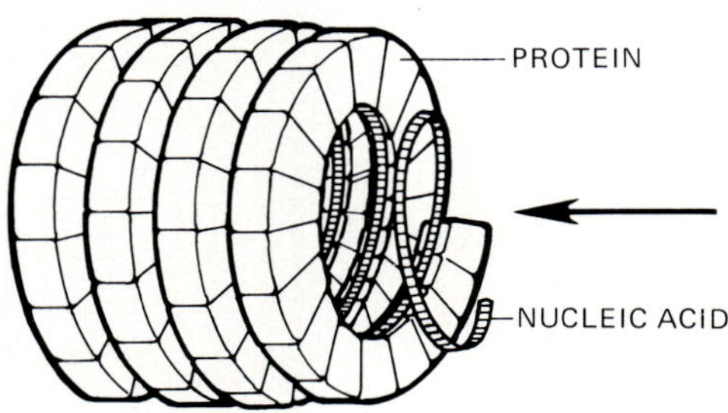

FIGURE 3.   Schematic diagram illustrating helical symmetry. The single rotational axis is indicated by an arrow. Protein subunits surround the nucleic acid, a portion of which is shown free in the drawing. The protein subunits of each turn of the helix are offset in relation to those of the previous turn.

Chapter 4

CLINICAL NEGATIVE STAIN ELECTRON MICROSCOPY

Basis for Use of Electron Microscopy in Diagnostic Virology

Viruses which can currently be identified by direct EM examination of body fluids or cellular debris are poxviruses, herpesviruses, hepatitis B-associated antigen, papovaviruses, reovirus in stools, some adenoviruses and a variety of viruses which are associated with gastroenteritis. Other viruses such as hepatitis A and coronaviruses can also be detected by immune EM. In this technique, specific antibody is used to aggregate the virus particles and thereby make identification easier. EM may also be a useful adjunct to conventional isolation and identification procedures for a wide variety of viruses, because any virus which can be propagated to about 10^6 particles/mℓ in tissue culture can easily be grouped by direct EM identification. Immune EM has also made it possible to type some isolates, but the technique is not yet widely used.

Poxviruses and herpesviruses can easily be discerned by direct EM of vesicular fluids or scrapings from lesions. In cases of overt infection, these viruses are present in lesions in sufficient numbers for detection by direct EM. They must be typed by conventional serological tests, however, because immune EM procedures for typing either pox viruses or herpesvirus isolates are not adequate for unequivocal identification.

More recently it has been found that cytomegalovirus can be detected in urine by direct EM, provided that the pseudoreplica technique is used to prepare the grid. This technique concentrates samples 30 to 40 times with a concomitant dialysis. It appears that in cytomegalic inclusion disease the virus is present in the urine in higher concentrations than indicated by infectivity titrations. Confirmation of cytomegalovirus is possible by immune EM, but techniques for this procedure have not been thoroughly evaluated.

Hepatitis B-associated antigen may be present in positive serum in large numbers. These particles can be detected by centrifugation of serum to pellet the virus and then by direct EM examination of the pelleted material. Hepatitis A has been seen by direct EM after purification from stools and has been detected in stools by immune EM.

Among the Papovaviridae, BK virus has been seen in urine samples from persons with organ transplants, JC virus has been seen in specimens of brain material taken at autopsy from patients with progressive multifocal leukoencephalopathy, and papilloma virus can easily be seen in samples from wart material. Further, simian vacuolating virus (SV40) can often be detected in tissue culture when CPE caused by this virus interferes with viral isolation attempts undertaken in monkey kidney tissue culture. Papovaviruses have a very distinctive morphology and may be found among cellular debris more easily than some other viruses.

A common use of EM in diagnostic virology is to detect viruses which cause gastroenteritis. Rotavirus is now known to be the major cause of gastroenteritis among young children, especially in the winter months. The virus is frequently present in feces in concentrations greater than 10^{10} particles/g and is easily detected by direct EM. Rotavirus has a very distinctive morphology which makes it easy to recognize. Infrequently, reovirus, which is sometimes found in stools, may be confused with rotavirus if the former particles are distorted when specimens are mishandled.

A variety of other viruses have also been associated with gastroenteritis, and are sometimes present in numbers sufficient for detection by EM. These include corona-like viruses, parvo-like viruses (Norwalk agent and similar agents), "astroviruses," caliciviruses, and "small round viruses." Adenoviruses are also occasionally found. When present in stools, the adenoviruses are generally in large numbers, but their significance has not been fully evaluated. Other gastroenteritis viruses will, no doubt, be found in the future.

One point of importance in EM examination of stools for viruses causing gastroenteritis is that these viruses are generally very stable. With the exception of the corona-like particles, the viruses mentioned above are stable to enzymatic digestion and lipid solvents which means that they can be extracted with solvents such as genetron. This treatment frees particles from most of their environmental debris and makes immune EM or direct EM more feasible.

Methodology

Specimens for diagnostic EM may be used directly for grid preparation if they are fluids. Vesicle fluid, urine, and tissue culture medium are examples of specimens which may be examined directly. Specimens such as feces must be suspended in buffered saline and appropriately diluted before grid preparation to prevent obscuring the entire grid with dense material. Tissue specimens may be macerated in a small amount of buffered saline and the supernatant fluid used for grids.

In the preparation of these viral specimens for EM, the technique of negative staining enhances the contrast of viral images that would otherwise be unimpressive. In a simple negative-stain drop technique, particles are mixed with a staining solution of a salt which is opaque to electrons. Grids coated with carbon or films such as Formvar® are floated on a drop of the mixture for a few seconds. They are then picked up and allowed to dry. In a similar "drop-to-drop" technique, the coated grid is first floated for a few seconds on a drop of the specimen fluid, blotted slightly by touching the edge with filter paper, then floated on a drop of stain. After a few seconds, the grid is again blotted slightly and air dried. Virus will be embedded in residual stain after drying. Particles and parts of particles not penetrated by the salt are seen as electron-lucent against an opaque background. Salt will penetrate between protruding parts and make them visible. Contrast is thus inverse (negative stain) because thin areas of stain appear electron-lucent and thick areas are electron-dense.

Some viral components such as nucleic acids will adsorb specific salts and therefore become positively stained. Thus uranyl acetate (UA) will positively stain nucleic acid and, indeed, one can often see positive and negative staining on different areas of the same grid when using this stain.

Many of the viruses in this atlas were prepared for electron microscopy by the pseudoreplica technique for negative-contrast staining. A drop of virus suspension is allowed to dry on a small block of 2.0% agarose and is then overlaid with two drops of 0.5% Formvar® in ethylene dichloride. The excess solvent is drained off, and the resulting thin Formvar® film is allowed to dry. The agar block is submerged in a small dish of stain, and the film is floated off onto the surface of the stain. A copper grid is placed onto the floating Formvar® film which contains adherent viral material. The grid is retrieved by a metal peg, and excess stain is removed by touching the membrane with filter paper. The grid is then ready for examination by EM. The pseudoreplica technique effectively removes salts from the virus preparations by dialysis into the agarose and thereby prevents formation of salt crystals and granules on the specimen grid. The stains used for these micrographs are either 0.5% aqueous UA or 2.0% phosphotungstic acid (PTA) buffered to pH 6.5 with KOH. These stains, composed of heavy metal salts, give good contrast with most virus preparations and aid in the preservation of virus structure during drying. A schematic of the pseudoreplica technique is outlined in Plate 1.

In the technique of immune electron microscopy, a viral sample is incubated with pre- and postinfection sera or with antiserum of known specificity. The specific antibody causes the virus particles to aggregate; when these particles are viewed by EM they appear in clumps, often obscured by a "fuzzy" coating of antibody. Various dilutions of the serum may be necessary to produce an easily discernable aggregation. Incubation is generally 1 hr at 37°C or overnight at 4°C. Grids may then be prepared by a drop technique or by pseudoreplication.

For magnification determinations in these micrographs, bovine catalase crystals (Worthington Biochemicals, Freehold, NJ) were used as an internal calibration standard. In some instances, the crystals were included with the virus sample, negatively stained, and photographed in the same field.

Determination of magnification was based on the method described by Wrigley.[2] The mean periodicity of the principal lattice spacing of the crystals is taken as 172 ± 4 Å. Half of this value (86 ± 2 Å), as represented by the space between the centers of each pair of electron-lucent vertical lines shown in Plate 2, is used in calculations to determine magnification. Measurements of the lattice spacing should be made on the original negatives and should be made across 50 to 100 repeats in an area where the lines are parallel. From this measurement and the known value of 86 Å, an accurate magnification value can be calculated.

REFERENCES

1. **Smith, K. O.**, Identification of viruses by electron microscopy, in *Methods in Cancer Research*, Busch, H., Ed., Academic Press, New York, 1967, 545.
2. **Wrigley, N. G.**, The lattice spacing of crystalline catalase as an internal standard of length in electron microscopy, *J. Ultrastruct. Res.*, 24, 454, 1968.

PSEUDOREPLICA TECHNIQUE FOR PREPARATION AND NEGATIVE STAINING OF VIRAL SPECIMENS FOR ELECTRON MICROSCOPY

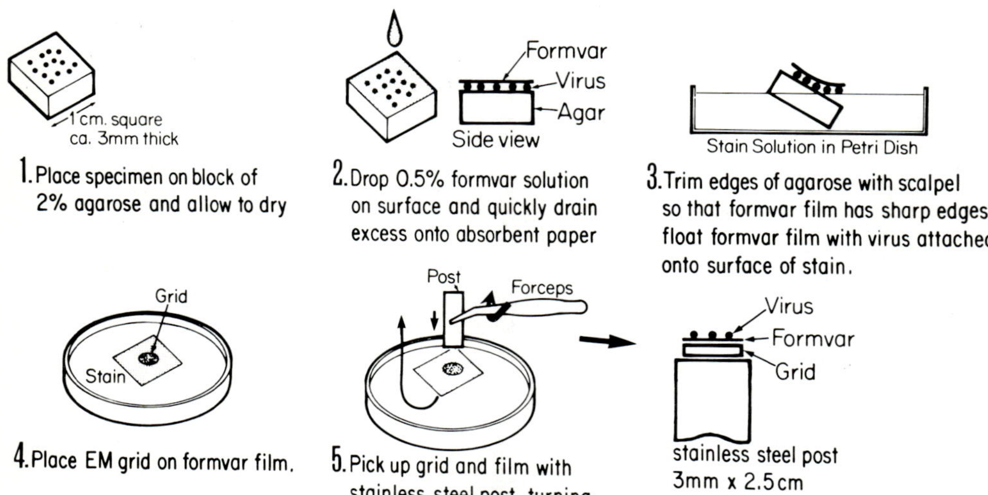

PLATE 1. A schematic of the pseudoreplica technique. A full description of the methodology and use of this technique has been reviewed by K. O. Smith (*Methods in Cancer Research,* Vol. 1, Busch, H., Ed., Academic Press, New York, 1967).

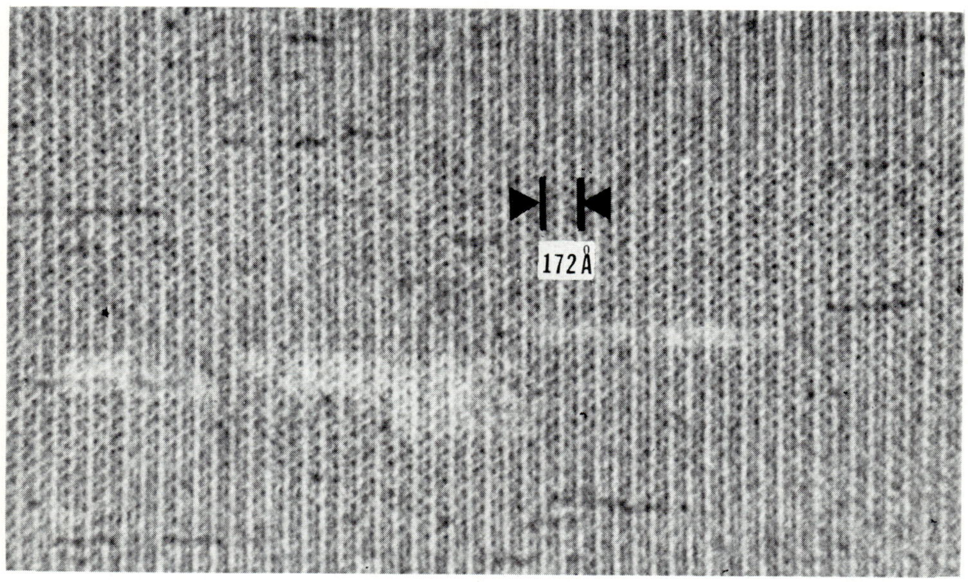

PLATE 2. Portion of a bovine catalase crystal negatively stained with UA. Marker indicates one complete lattice spacing, 172 ± 4 Å.

Part I
RNA Viruses with
Cubic Capsid Symmetry
and Naked Nucleocapsids

Chapter 5

PICORNAVIRIDAE

The name picornavirus (pico = small; rna = ribonucleic acid genome) was introduced to designate a group including enteroviruses (enteric) and rhinoviruses (rhino = nose) because of fundamental similarities in their various biological and biochemical properties. Caliciviruses (calicle = cup-like depression) were later added, and the group was given family status. The family Picornaviridae now includes three genera of animal viruses: *Enterovirus, Rhinovirus,* and *Calicivirus.*

Enteroviruses cause diseases in humans ranging from severe paralysis (poliovirus) to minor respiratory or undifferentiated febrile illnesses. The Coxsackie viruses produce such diseases as aseptic meningitis and pericarditis, while ECHO viruses cause aseptic meningitis and common colds. The rhinoviruses, with over a hundred members are, however, the major cause of the common cold. Foot-and-mouth disease virus is a rhinovirus of exceptional veterinary importance. Caliciviruses have been found in swine, seals, cats, and, more recently, humans. The virus from humans has been recovered from fecal material of persons ill with gastroenteritis, but no causal relationship has yet been firmly established.

Enteroviruses and rhinoviruses are naked icosahedral viruses 20 to 30 nm in diameter. All have the same appearance when viewed by negative-stain EM. They have capsomeric symmetry which has not been satisfactorily delineated. There is evidence, however, that there are 32 capsomeres arrayed in a T = 3 arrangement. Caliciviruses are larger (35 to 40 nm) than other picornaviruses and have a more definitive surface structure. Cup-like surface depressions are prominent and are sometimes seen in a sixfold "Star of David" array. The morphology of caliciviruses has not been firmly established and several models have been proposed. The most recent evidence suggests a virion with trimeric clustering of 180 polypeptides in a T = 3 lattice, but this has not been definitely confirmed. In view of these striking differences, this group is to be split off from the Picornaviridae to become a new virus family — Caliciviridae.

Because of their small size the picornaviruses are some of the most difficult of the mammalian viruses to detect by negative-stain EM, either in body fluids or tissue culture.

REFERENCES

1. **Almeida, J. D., Waterson, A. P., Prydie, J., and Fletcher, E. W. L.,** The structure of a feline picornavirus and its relevance to cubic viruses in general, *Arch. Gesamte Virusforsch.,* 25, 105, 1968.
2. **Baltimore, D.,** The replication of picornaviruses, in *Biochemistry of Viruses,* Levy, H. B., Ed., Marcel Dekker, New York, 1969, 101.
3. **Burroughs, J. N., Doel, T. R., Smale, C. J., and Brown, F.,** A model for vesicular exanthema virus, the prototype of the calicivirus group, *J. Gen. Virol.,* 40, 161, 1978.
4. **Dalldorf, G.,** The Coxsackie viruses, *Ann. Rev. Microbiol.,* 9, 277, 1955.
5. **Horne, R. W. and Nagington, J.,** Electron microscope studies of the development and structure of poliomyelitis virus, *J. Mol. Biol.,* 1, 333, 1959.
6. **Levintow, L.,** Reproduction of picornaviruses, in *Comprehensive Virology,* Vol. 2, Fraenkel-Conrat, H. and Wagner, R. R., Eds., Plenum Press, New York, 1974, 109.
7. **Mayor, H.,** Picornavirus symmetry, *Virology,* 22, 156, 1964.
8. **McGregor, S. and Mayor, H.,** Biophysical studies on rhinovirus and poliovirus. I. Morphology of viral ribonucleoprotein, *J. Virol.,* 2, 149, 1968.
9. **McGregor, S., Hall, L., and Rueckert, R. R.,** Evidence for the existence of protomers in the assembly of encephalomyocarditis virus, *J. Virol.,* 15, 1107, 1975.

10. **Peterson, J. E. and Studdert, M. J.,** Feline picornavirus. Structure of the virus and electron microscopic observations on infected cell cultures, *Arch. Gesamte Virusforsch.,* 32, 249, 1970.
11. **Phillips, B. A. and Fennel, R.,** Polypeptide composition of poliovirions, naturally occurring empty capsids, and 14S precursor particles, *J. Virol.,* 12, 291, 1973.
12. **Rueckert, R. R.,** Picornaviral architecture, in *Comparative Virology,* Maramorosch, K. and Kurstak, E., Eds., Academic Press, New York, 1971, 255.
13. **Rueckert, R. R.,** On the structure and morphogenesis of picornaviruses, in *Comprehensive Virology,* Vol. 6, Fraenkel-Conrat, H. and Wagner, R. R., Eds., Plenum Press, New York, 1976, 131.
14. **Schaffer, F. L.,** Caliciviruses, in *Comprehensive Virology,* Vol. 14, Fraenkel-Conrat, H. and Wagner, R. R., Eds., Plenum Press, New York, 1979, 249.
15. **Tyrrell, D. A. J.,** Rhinoviruses, *Virol. Monogr.,* 2, 68, 1968.
16. **Wenner, H. A. and Behbehani, A. M.,** Echoviruses, *Virol. Monogr.,* 1, 1, 1968.
17. **Zwillenberg, L. O. and Burki, F.,** On the capsid structure of some small feline and bovine RNA viruses, *Arch. Gesamte Virusforsch.,* 19, 373, 1966.

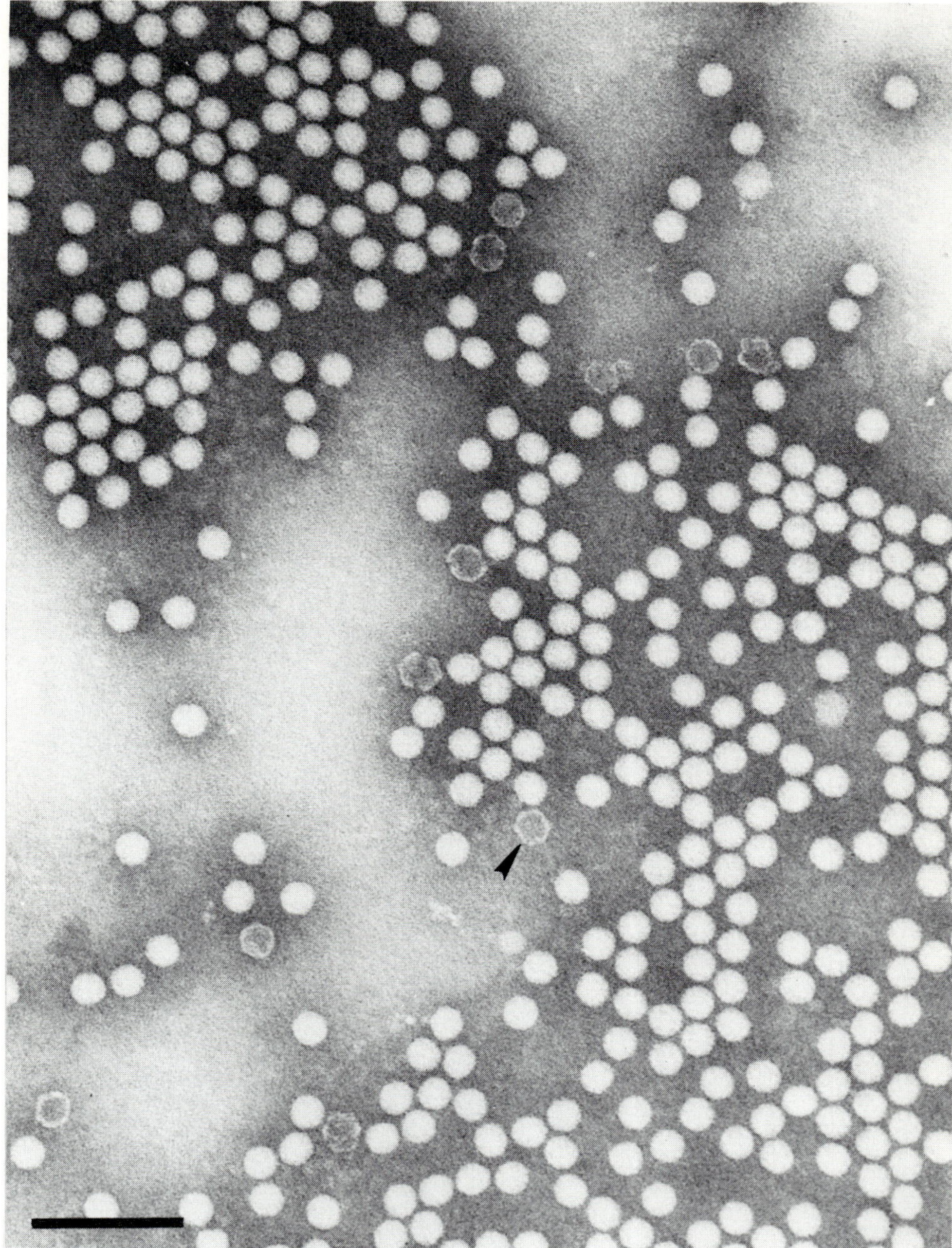

PLATE 1. *Poliovirus* type 1. Virus was purified by CsCl density gradient centrifigation. Icosahedral symmetry is evident but capsomeric arrangement can not be definitely delineated because the viruses are very small (20 to 30 nm). The arrow points to a particle which is "empty" in that it has no RNA genome. All bars equal 100 nm.

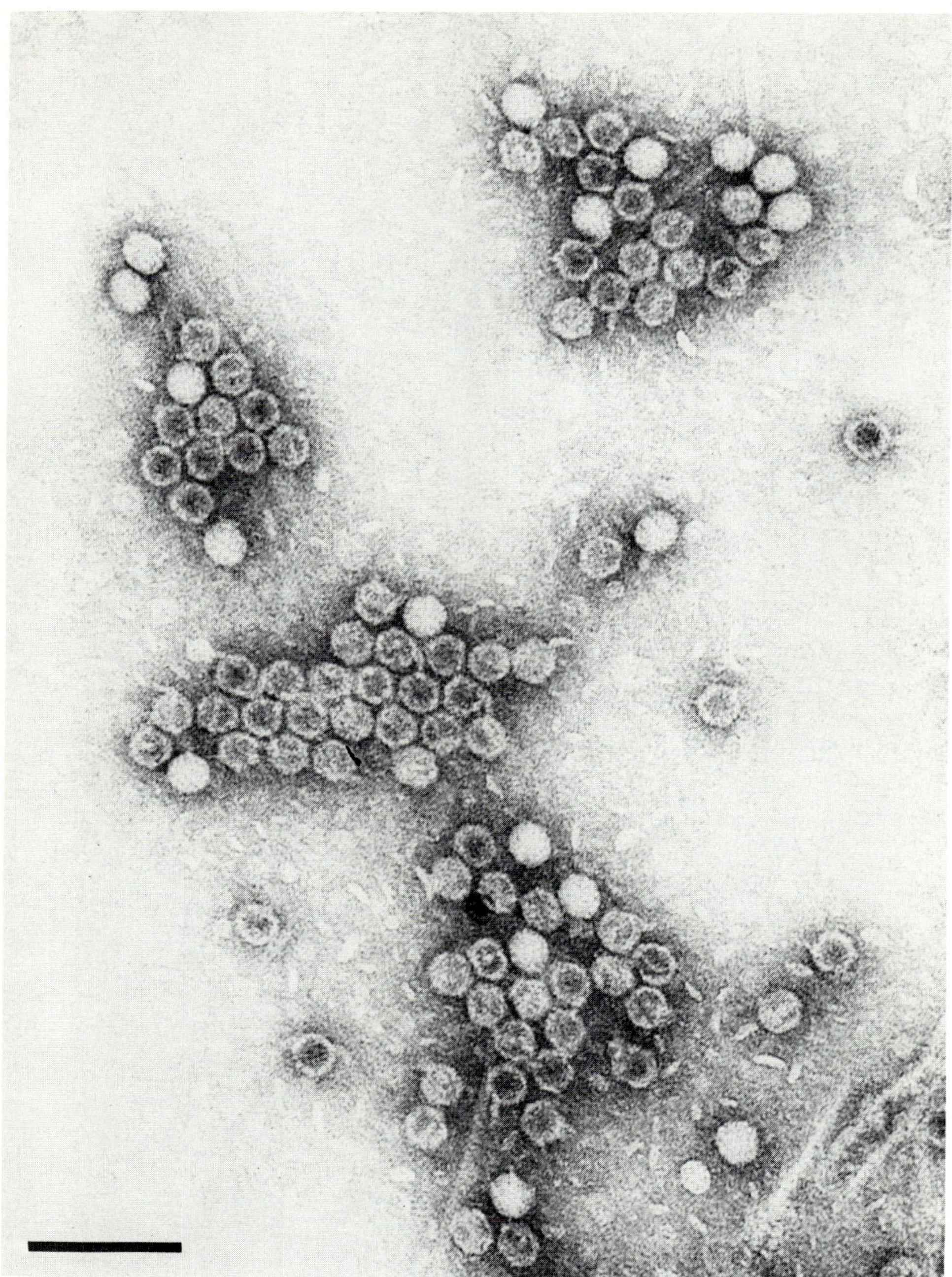

PLATE 2. Empty poliovirus in crystalline array.

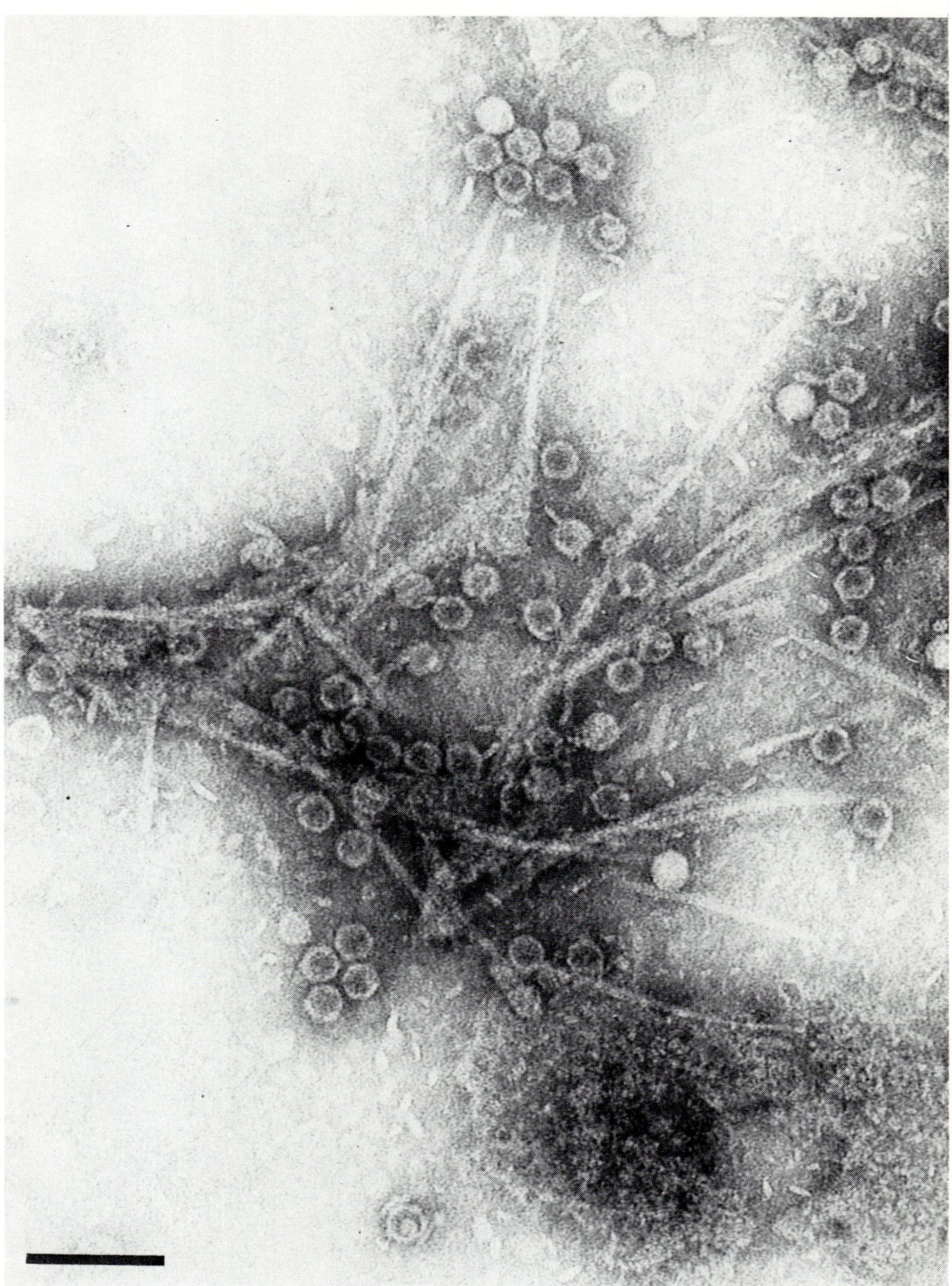

PLATE 3. Poliovirus disrupted by alkaline shock.

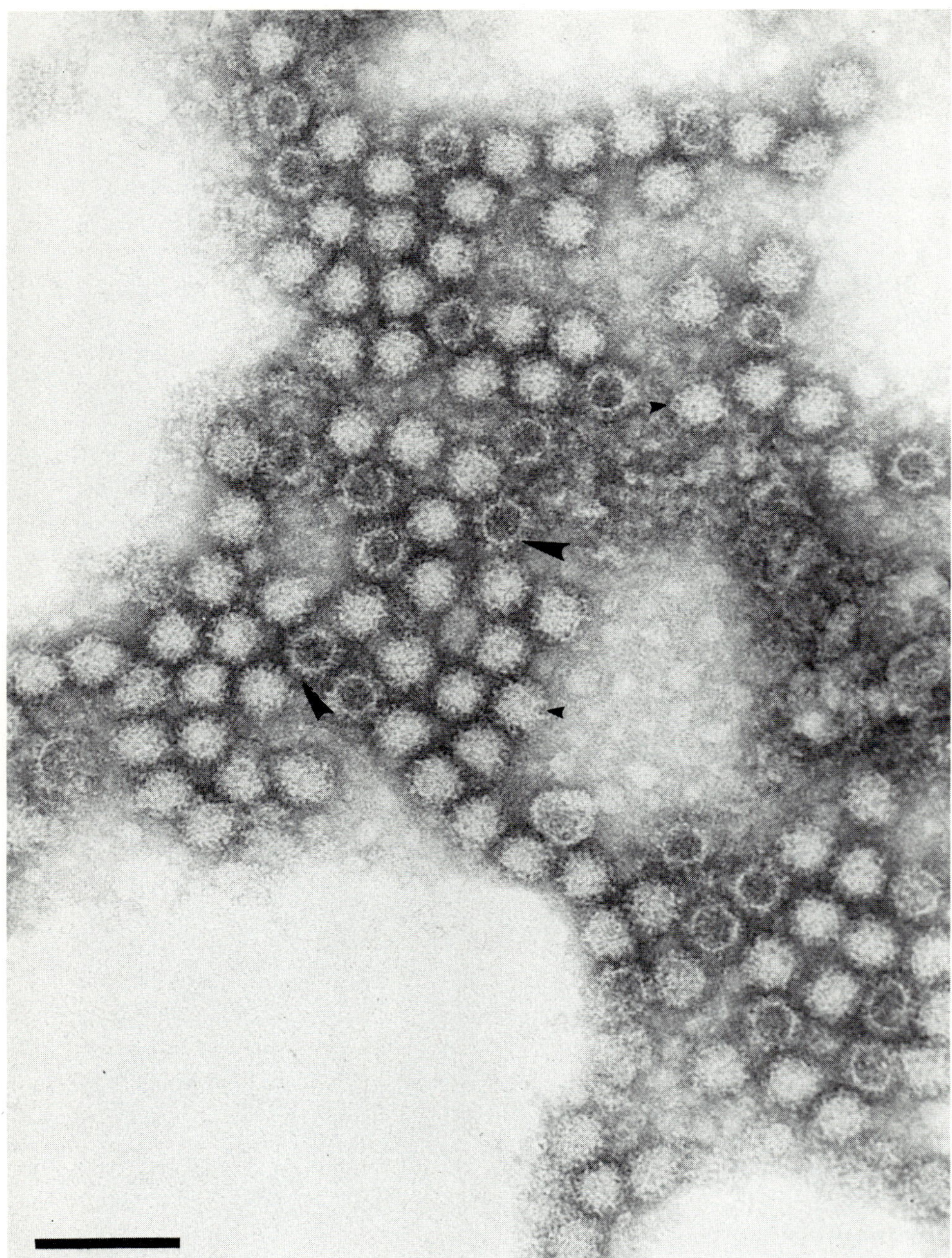

PLATE 4. *Calicivirus* (feline). Caliciviruses are slightly larger than other viruses of the Picornaviridae. They average 35 to 40 nm in diameter. These viruses have distinctive cup-like surface depressions (short arrows) which are sometimes seen in a ''Star of David'' array. Electron-dense particles are empty and the core is outlined by a rim. Well-spaced projections (long arrows) radiate from the rim.

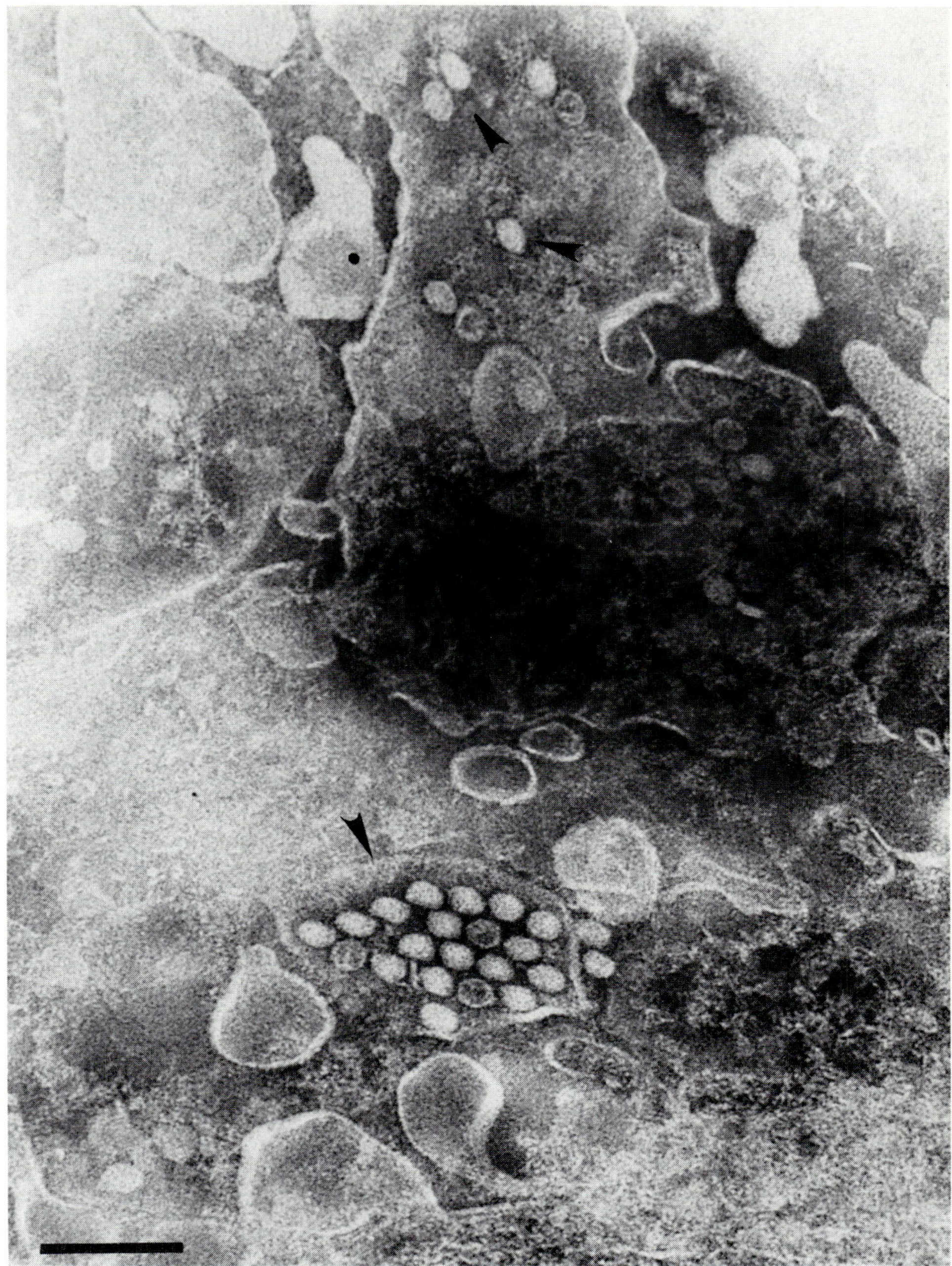

PLATE 5. Enterovirus isolated from eye fluid of a person with conjunctivitis. Virus was propagated in tissue culture and culture fluids were examined by negative stain EM. A group of particles is seen in a cell vesicle, and free-lying particles are also evident (arrows).

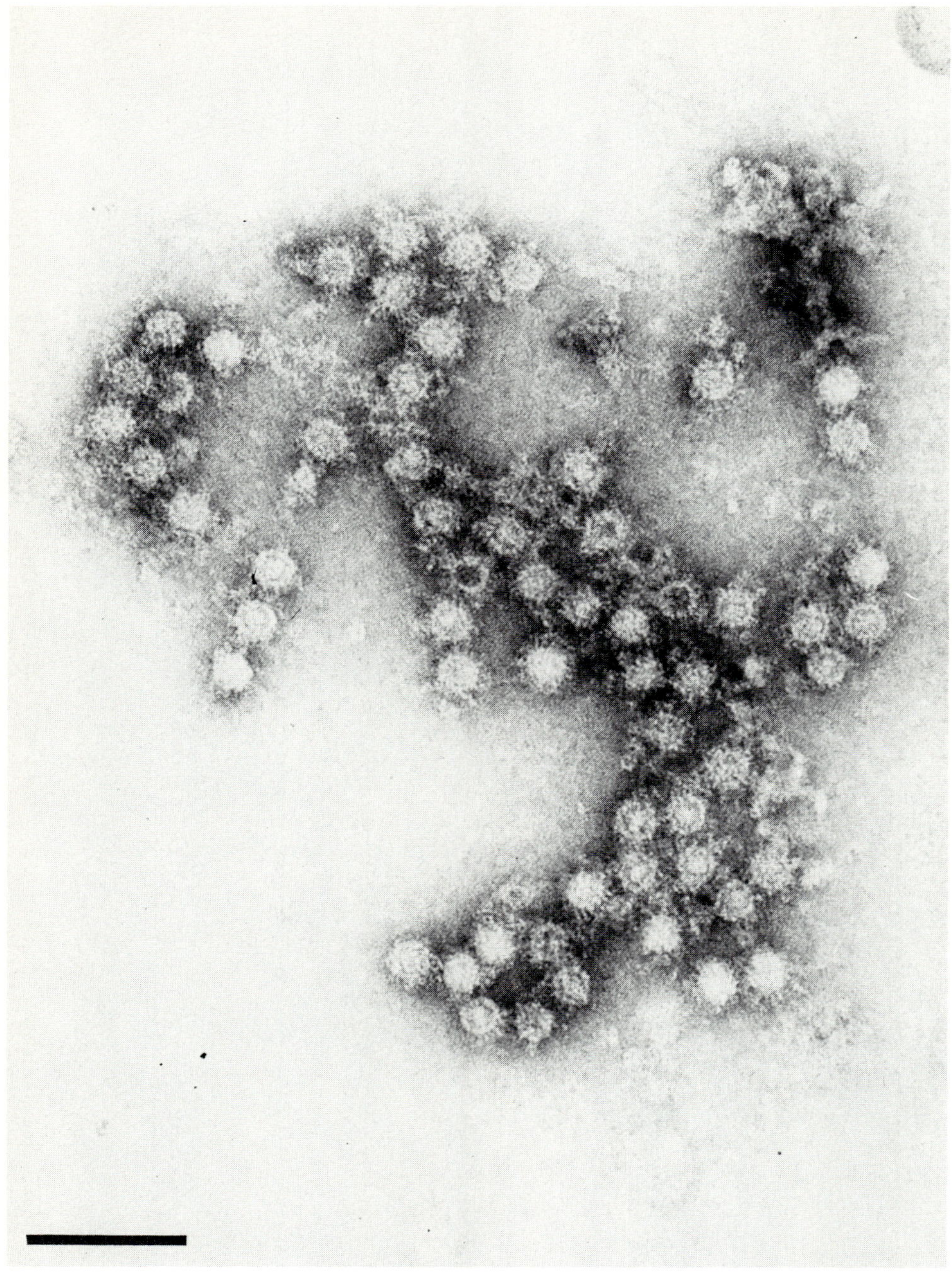

PLATE 6. Immune electronmicroscopy of Coxsackie virus. Antibody on the surface of clumped particles is clearly evident.

Chapter 6

REOVIRIDAE

There are currently three recognized genera of the family Reoviridae which infect vertebrates. These are *Reovirus, Orbivirus,* and *Rotavirus.* Member viruses are distinctive in that they possess two clearly defined capsid shells surrounding double-stranded RNA genomes which consist of 10 or more segments. They are the only vertebrate viruses known to have a capsid formed by shared subunits as a prominent feature. The members of the three genera are not antigenically related, but viruses within each genus show extensive antigenic cross-reactivity.

Members of the *Reovirus* genus, represented by types 1, 2, and 3, are widely distributed in nature and have been recovered worldwide from the respiratory and intestinal tracts of humans and from various lower animals. From humans reovirus has been recovered both from healthy persons and persons with a wide variety of illnesses. However, no specific illness has been definitely associated with any of the reoviruses of vertebrates (thus the name Reo; r, respiratory; e, enteric; o, orphan).

Complete reovirus particles are 75 nm in diameter. Our studies indicate that the overall architecture of the outer capsid is that of a T = 3, with probably 32 large capsomeres, which themselves have the more complex symmetry of a T = 9 icosadeltahedron. The units do not exhibit clustering with wide stain-filled spaces between them, as do isometric viruses such as herpesviruses or adenoviruses. Instead, the juxtaposition of morphological units results in an array of holes. Other investigators have proposed that the capsid has 180, 92, and 127 capsomeres. The outer capsid layer of the virion can be removed by enzymatic treatment. The resulting particle (inner capsid) surrounding the core is an icosahedron about 45 nm in diameter which has surface capsomeres and projections at the 12 vertices. They extend about 5 nm from the core surface, are 10 nm in diameter and have a central channel 5 nm in diameter.

Members of the *Orbivirus* genus are transmitted by arthropods and are thus considered to be arboviruses. They were included in the Reoviridae family when it was established that, like reoviruses, they had two capsid layers and possessed double-stranded segmented RNA genomes. Bluetongue virus of sheep (transmitted by *Culicoides* gnats) and Colorado tick fever virus (transmitted by ticks) are examples of orbiviruses which cause veterinary and human health problems, respectively. All intact virions have a structurally featureless outer layer, and particles vary from 60 to 80 nm in diameter by negative-stain EM. The outer layer may spontaneously disintegrate when virus preparations are subjected to laboratory manipulation. The freed inner capsid can then be seen as a shell of 32 large ring-shaped capsomeres (Orbi = ring) which surrounds a 38-nm icosahedral core. The inner capsid of bluetongue virus is 56 nm in diameter. The capsomeres are hollow structures 10 to 12 nm wide with a 4-nm axial hole and are arrayed with a T = 3 primary symmetry and a more complicated secondary structure. Free-lying cores are seen on axes of fivefold, threefold, and twofold symmetry typical of icosahedral particles.

The genus *Rotavirus* comprises an immunologically related group of viruses that causes enteritis in many mammals, including man. Human rotavirus is the major cause of enteritis of young children throughout the world, and antigenically related rotaviruses cause gastroenteritis in the young of a wide variety of mammals; one is the cause of calf scours disease. Virus can be detected in fecal extracts by direct negative-stain electron microscopy. Intact rotaviruses are 67 nm in diameter and have a distinctly characteristic outer rim. Capsomeres from the inner layer radiate toward the rim, giving the virus the appearance of a wheel (Rota = wheel).

Particles without the outer rim layer are 58 nm in diameter and consist of a shell of large morphological units arrayed with a complex primary and secondary symmetry. As with the outer shell of reoviruses and the inner shell of orbiviruses, these units are formed by sharing of trimeric subunits. The arrangement of the subunits has not been definitely established. We have concluded that there are 32 large morphological units arrayed into a T = 3 primary symmetry, with a more complex secondary symmetry of a T = 9 icosadeltahedron. Others have interpreted the symmetry as T = 9, T = 16, and the skew, T = 13. The inner shell surrounds an icosahedral core 38 nm in diameter. The inner capsid and core of rotaviruses and orbiviruses appear to be identical in morphology. They differ from reoviruses in that the inner capsids of the other genera do not have projections from the vertices as do reoviruses.

REFERENCES

Reovirus

1. **Amano, Y., Katagiri, S., Ishida, N., and Watanabe, Y.,** Spontaneous degradation of reovirus capsid into subunits, *J. Virol.,* 8, 805, 1971.
2. **Dales, S., Gomatos, P., and Hsu, K. C.,** The uptake and development of reovirus in strain L cells followed with labeled viral ribonucleic acid and ferritin-antibody conjugates, *Virology,* 25, 193, 1965.
3. **Gauntt, C. J. and Graham, A. F.,** The reoviruses, in *The Biochemistry of Viruses,* Levy, H. B., Ed., Marcel Dekker, New York, 1969, 259.
4. **Gomatos, P. J. and Tamm, I.,** The secondary structure of reovirus RNA, *Proc. Natl. Acad. Sci. USA,* 49, 707, 1963.
5. **Gomatos, P. J., Tamm, I., Dales, S., and Franklin, R. M.,** Reovirus type 3: physical characteristics and interaction with L cells, *Virology,* 17, 441, 1962.
6. **Joklik, W. K.,** The molecular biology of reovirus, *J. Cell Physiol.,* 76, 289, 1970.
7. **Joklik, W. K.,** Reproduction of Reoviridae, in *Comprehensive Virology,* Vol. 2, Fraenkel-Conrat, H. and Wagner, R. R., Eds., Plenum Press, New York, 1974, 231.
8. **Jordan, L. E. and Mayor, H. D.,** The fine structure of reovirus, a new member of the icosahedral series, *Virology,* 17, 597, 1962.
9. **Loh, P. C., Hohl, H. R., and Soergel, M.,** Fine structure of reovirus type 2, *J. Bacteriol.,* 89, 1140, 1965.
10. **Luftig, R. B., Kilham, S., Hay, A. J., Zweerink, H. J., and Joklik, W. K.,** An ultrastructural study of virions and cores of reovirus type 3, *Virology,* 48, 170, 1972.
11. **Mayor, H. D., Jamison, R. M., Jordan, L. E., and Van Mitchell, M.,** Reoviruses. II. Structure and composition of the virion, *J. Bacteriol.,* 89, 1548, 1965.
12. **Mayor, H. D. and Jordan, L. E.,** Preparation and properties of the internal capsid components of reovirus, *J. Gen. Virol.,* 3, 233, 1968.
13. **Palmer, E. L. and Martin, M. L.,** The fine structure of the capsid of reovirus type 3, *Virology,* 76, 109, 1977.
14. **Rosen, L.,** Reoviruses, *Virology Monogr.,* 1, 73, 1968.
15. **Shatkin, A. J.,** Viruses containing double-stranded RNA, in *Molecular Basis of Virology,* Fraenkel-Conrat, H. Ed., Reinhold, New York, 1968, 351.
16. **Smith, R. E., Zweerink, H. J., and Joklik, W. K.,** Polypeptide components of virions, top component and cores of reovirus type 3, *Virology,* 39, 791, 1969.
17. **Vasquez, C. and Tournier, P.,** The morphology of reovirus, *Virology,* 17, 503, 1962.
18. **Vasquez, C. and Tournier, P.,** New interpretation of reovirus structure, *Virology,* 24, 128, 1964.

Orbivirus

19. **Borden, E. C., Shope, R. E., and Murphy, F. A.,** Physicochemical and morphological relationships of some arthropod-borne viruses to bluetongue virus — a new taxonomic group. Physicochemical and serological studies, *J. Gen. Virol.,* 13, 261, 1971.
20. **Bowne, J. G. and Ritchie, A. E.,** Some morphological features of bluetongue virus, *Virology,* 40, 903, 1970.
21. **Els, H. J. and Verwoerd, D. W.,** Morphology of bluetongue virus, *Virology,* 38, 213, 1969.

22. Howell, P. G. and Verwoerd, D. W., Bluetongue virus, *Virol. Monogr.*, 9, 37, 1971.
23. Martin, S. A. and Zweerink, H. J., Isolation and characterization of two types of bluetongue virus particles, *Virology*, 50, 495, 1972.
24. Murphy, F. A., Coleman, P. H., Hansen, A. K., and Gary, G. W. Jr., Colorado tick fever virus, an electron microscope study, *Virology*, 35, 28, 1968.
25. Murphy, F. A., Borden, E. C., Shope, R. E., and Harrison, A., Physicochemical and morphological relationships of some arthropod-borne viruses to bluetongue virus — a new taxonomic group. Electron microscopic studies, *J. Gen. Virol.*, 13, 273, 1971.
26. Oellermann, R. A., Els, H. J., and Erasmus, B. J., Characterization of African horse-sickness virus, *Arch. Gesamte Virusforsch.*, 29, 163, 1970.
27. Verwoerd, D. W., Els, H. J., DeVilliers, E. M., Huismans, H., Structure of the bluetongue virus capsid, *J. Virol.*, 10, 783, 1972.
28. Verwoerd, D. W., Huismans, H., and Erasmus, B. J., Orbiviruses, in *Comprehensive Virology*, Vol. 14, Fraenkel-Conrat, H. and Wagner, R. R., Eds., Plenum Press, New York, 1979, 285.

Rotavirus

29. Bishop, R. F., Davidson, G. P., Holmes, I. H., and Ruck, B., Virus particles in epithelial cells of duodenal mucosa from children with acute nonbacterial gastroenteritis, *Lancet*, 2, 1281, 1973.
30. Bishop, R. F., Davidson, G. P., Holmes, I. H., and Ruck, B., Detection of a new virus by electron microscopy of faecal extracts from children with acute gastroenteritis, *Lancet*, 1, 149, 1974.
31. Bridger, J. C. and Woode, G. N., Characterization of two particle types of calf rotavirus, *J. Gen. Virol.*, 31, 245, 1976.
32. Esparza, J. and Gil, F., A study on the ultrastructure of human rotavirus, *Virology*, 91, 141, 1978.
33. Flewett, T. H., Bryden, A. S., and Davies, H., Virus particles in gastroenteritis, *Lancet*, 2, 1497, 1973.
34. Flewett, T. H., Davies, H., Bryden, A. S., and Robertson, M. J., Diagnostic electron microscopy of faeces. II. Acute gastroenteritis associated with reovirus-like particles, *J. Clin. Pathol.*, 27, 608, 1974.
35. Holmes, I. H., Ruck, B. J., Bishop, R. F., and Davidson, G. P., Infantile enteritis viruses: morphogenesis and morphology, *J. Virol.*, 16, 937, 1974.
36. Kogasaka, R., Akihara, M., Horino, K., Chiba, S., and Nakao, T., A morphological study of human rotavirus, *Arch. Virol.*, 61, 41, 1979.
37. Martin, M. L., Palmer, E. L., and Middleton, P. J., Ultrastructure of infantile gastroenteritis virus, *Virology*, 68, 146, 1975.
38. Palmer, E. L., Martin, M. L., and Murphy, F. A., Morphology and stability of infantile gastroenteritis virus: comparison with reovirus and bluetongue virus, *J. Gen. Virol.*, 35, 403, 1977.
39. Roseto, A., Escaig, J., Delain, E., Cohen, J. and Scherrer, R., Structure of rotavirus as studied by the freeze-drying technique, *Virology*, 98, 471, 1979.
40. Stannard, L. M. and Schoub, B. D., Observations on the morphology of two rotaviruses, *J. Gen. Virol.*, 37, 435, 1977.
41. Wyatt, R. G., Kalica, A. R., Mebus, C. A., Kim, H. W., London, W. T., Chanock, R. M., and Kapikian, A. Z., Reovirus-like agents (rotaviruses) associated with diarrheal illness in animals and man, in *Perspectives in Virology*, Vol. 10, Pollard, M., Ed., Raven Press, New York, 1978, 121.

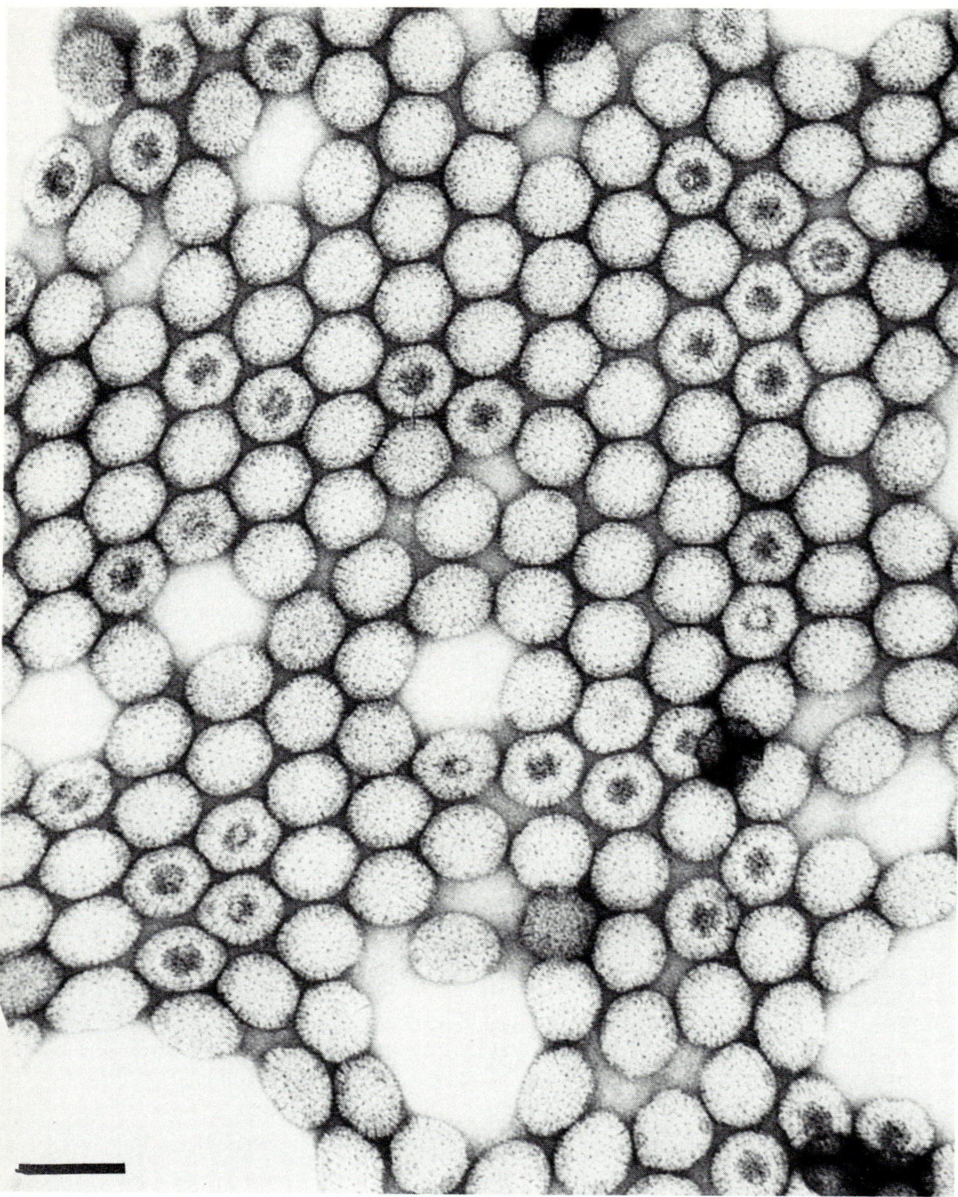

PLATE 1. Intact reovirus type 3. Both inner and outer capsid shells are present and an occasional particle can be seen to be penetrated by negative stain. All bars equal 100 nm.

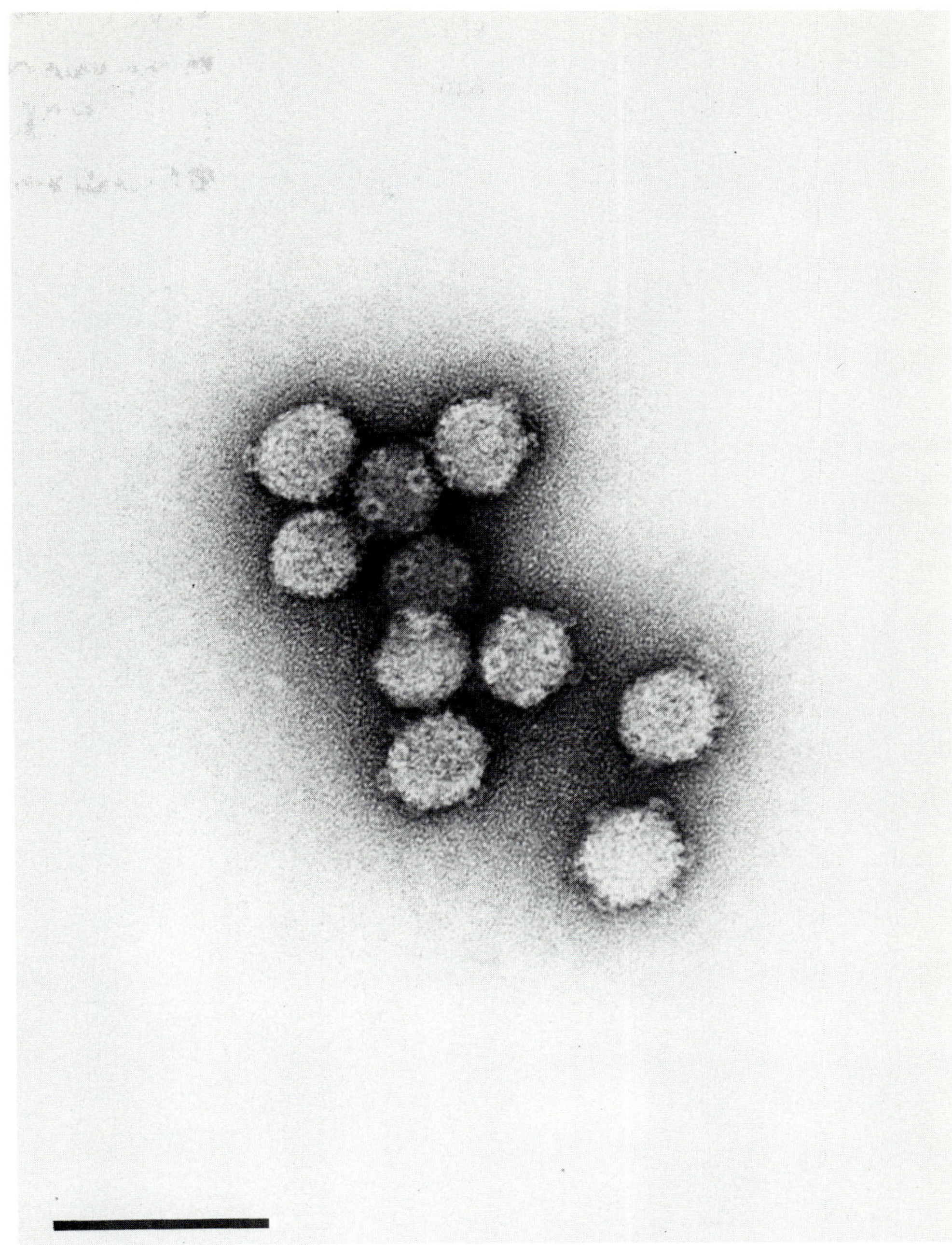

PLATE 2. Reovirus inner capsids obtained by enzymatic treatment of complete particles. Projections from vertices of the icosahedron are evident. There are 12 such projections on the surface of each subviral particle.

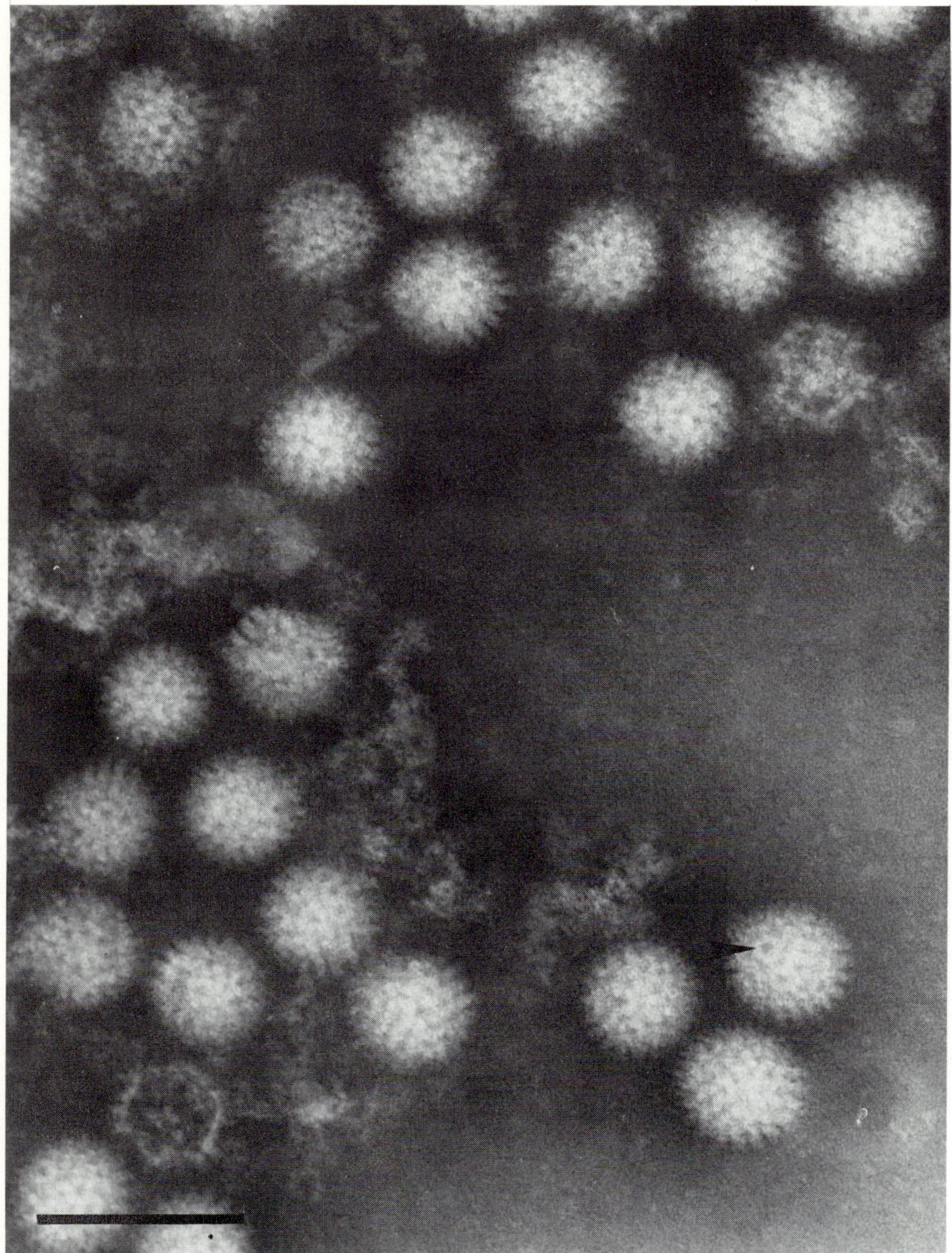

PLATE 3.   Bluetongue virus. The outer layer of intact particles is "hazy" in appearance. Ring-shaped capsomeres (arrow) can be seen on the surface of the inner capsid.

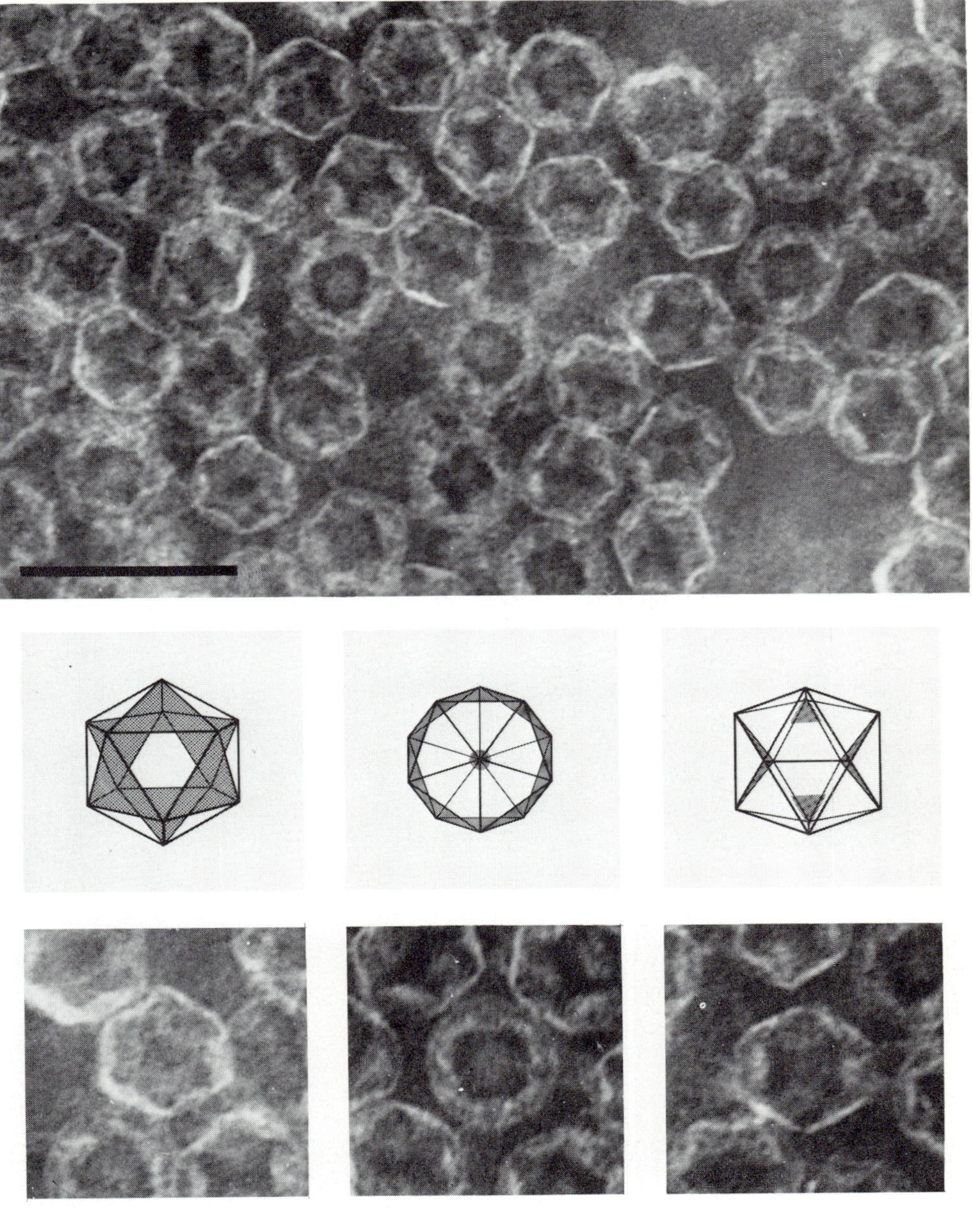

PLATE 4.   Icosahedral cores characteristic of both orbiviruses and rotaviruses. The photographs in the lower part of the plate show diagrams of three-, five-, and twofold axes of symmetry, together with orbivirus cores on corresponding axes of symmetry.

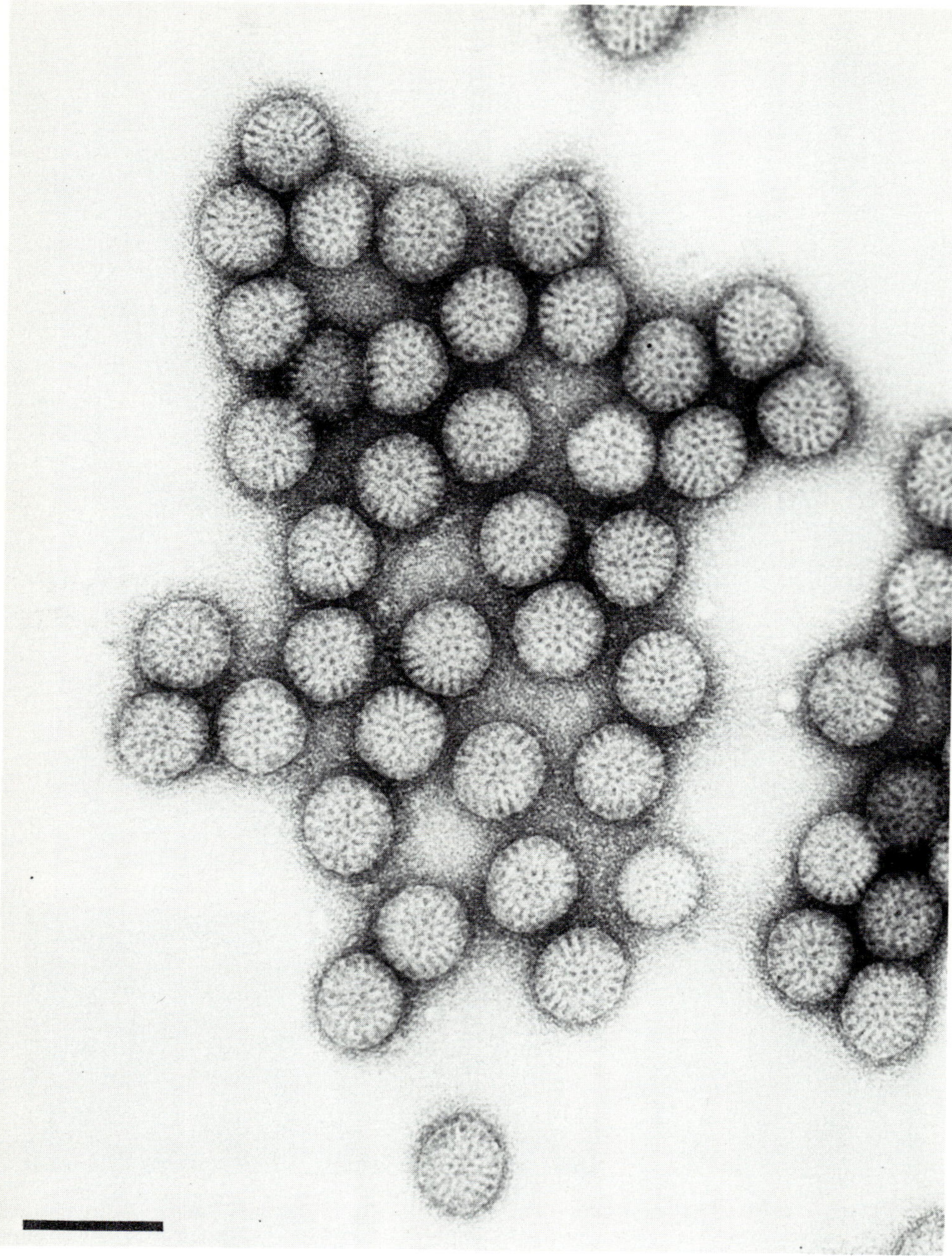

PLATE 5.   Intact rotavirus (double-shelled) particles. The distinctive rim and radiating capsomeres give particles a "wheel" shape.

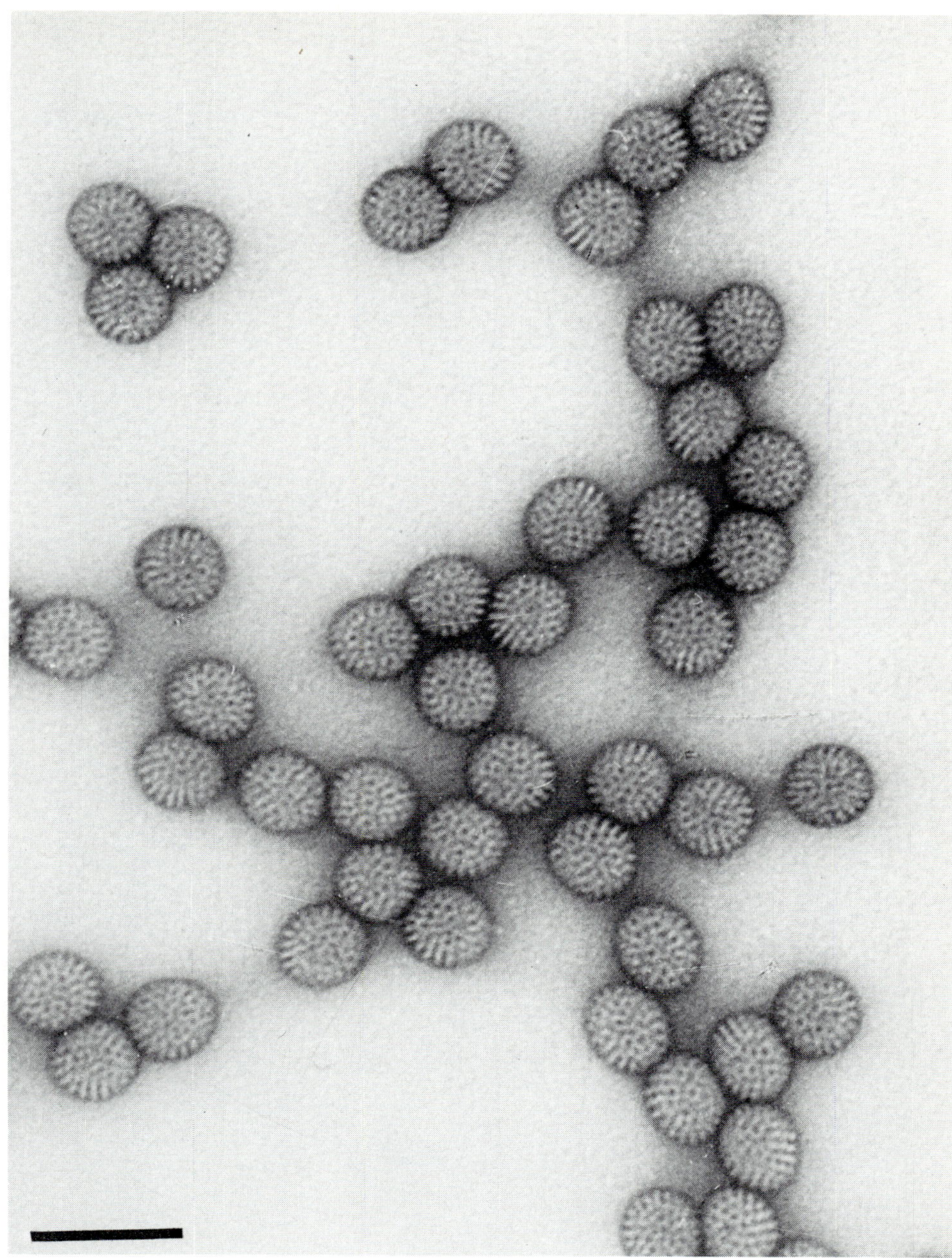

PLATE 6.   Inner capsid (single-shelled particles) of rotavirus. At low magnification when particles are in certain orientation, the capsomeres are seen as ring-shaped, the same as the inner capsid of orbiviruses.

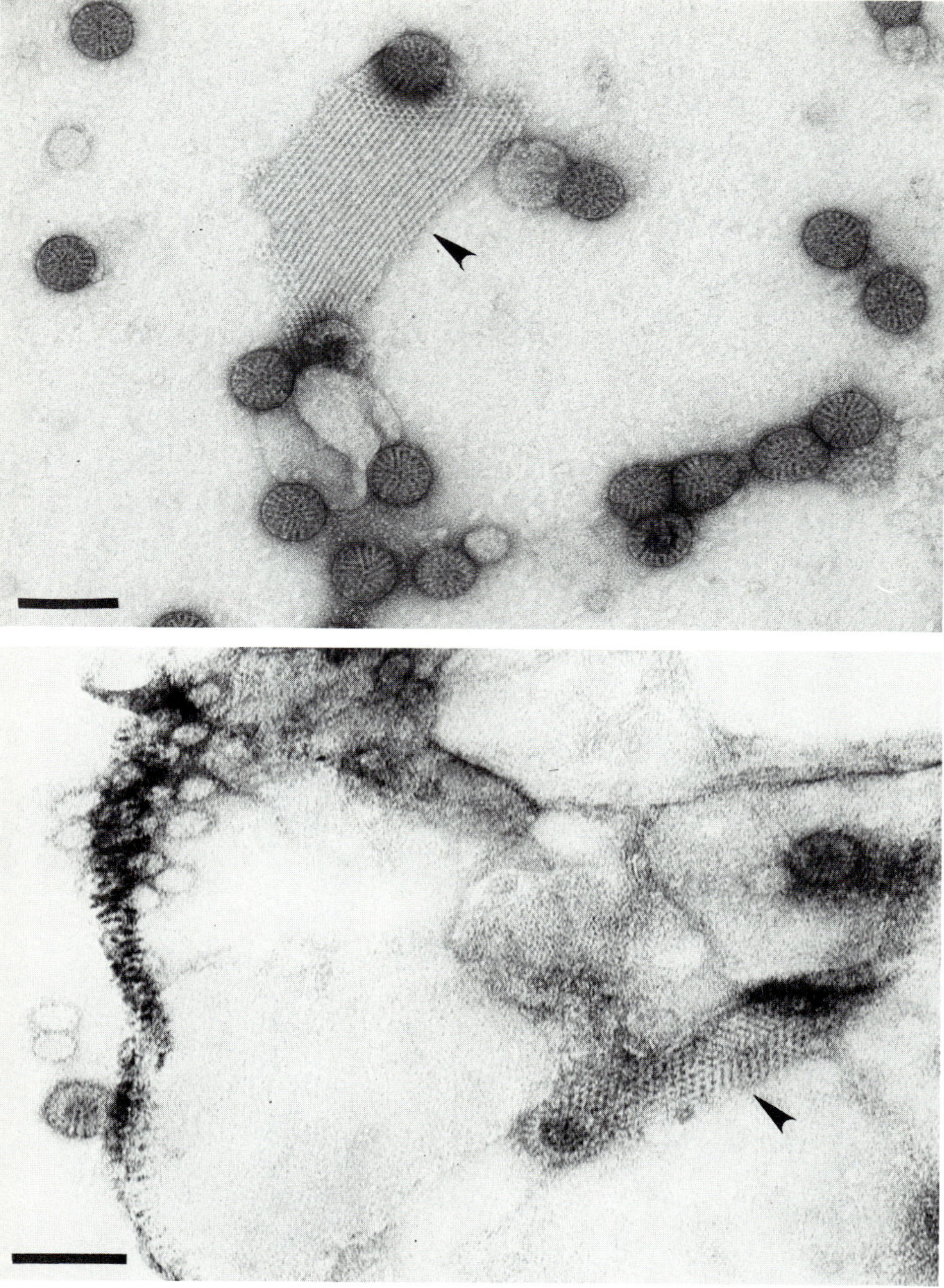

PLATE 7.   Rotavirus among debris of stool. These viruses are usually present in large numbers and are easily recognized. Lattice arrays (arrow) are commonly seen in such preparations.

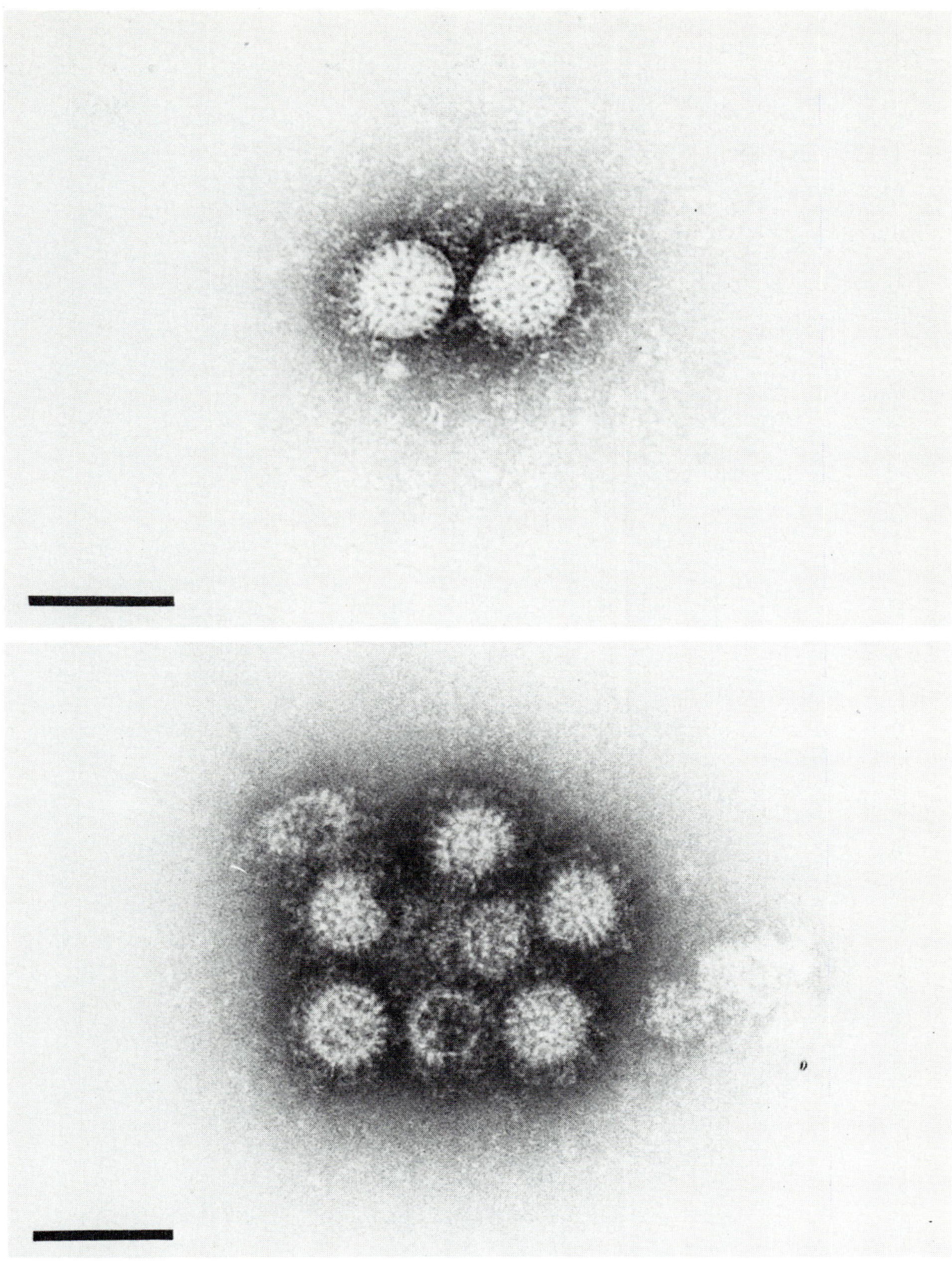

PLATE 8.   Immune EM of human rotavirus. In the top panel are double-shelled particles covered with colostral antibody. The bottom panel is a group of single-shelled particles aggregated by antiserum.

Part II
RNA Viruses with
Cubic Capsid Symmetry and Envelopes

Chapter 7

TOGAVIRIDAE

The Togaviridae (toga = gown, cloak) consists of four genera; *Alphavirus* containing the group A arboviruses, *Flavivirus* containing the group B arboviruses, *Rubivirus* with rubella as the type species, and *Pestivirus*, a genus of animal pathogens with mucosal-disease virus as the type species.

The *Alphavirus* genus contains species such as western and eastern encephalitis viruses which cause severe encephalitis in equine animals and humans. The *Flavivirus* genus contains viruses which cause a variety of maladies ranging from systemic disease such as yellow fever to severe neurologic disease such as St. Louis encephalitis and Japanese encephalitis to hemorrhagic fevers such as Kyasanur Forest disease. Rubella virus is the only species in the genus *Rubivirus.*

The Togaviridae are all morphologically similar. They are spherical and 40 to 70 nm in diameter. The only notable difference is that members of the *Alphavirus* genus are somewhat larger than those of the other genera. The togaviruses have a lipoprotein envelope tightly attached to an icosahedral nucleocapsid. The capsid of the alphavirus Sindbis appears to be composed of 32 capsomeres about 14 nm in diameter. The surface subunits are also apparently arrayed in an icosahedral surface lattice, possibly in a T = 4 configuration. Sindbis has been the only virus of the Togaviridae which has been studied in detail by negative-stain EM; even so, there is some disagreement concerning its structure. Visualization of togaviruses by negative stain EM has not been too satisfactory because particles tend to appear nondescript.

REFERENCES

1. **Acheson, N. H. and Tamm, I.,** Purification and properties of Semliki Forest virus nucleocapsids, *Virology,* 41, 306, 1970.
2. **Berge, T. O.,** International catalogue of arboviruses including certain other viruses of vertebrates, U.S. Dept. of HEW, DHEW Publication no. (CDC) 75-8301, 1975.
3. **Brown, D. T. and Gliedman, J. B.,** Morphological variants of Sindbis virus obtained from infected mosquito tissue culture cells, *J. Virol.,* 12, 1534, 1973.
4. **Brown, D. T., Waite, M. R. F., and Pfefferkorn, E. R.,** Morphology and morphogenesis of Sindbis virus as seen with freeze-etching techniques, *J. Virol.,* 10, 524, 1972.
5. **Enzmann, P. J. and Weiland, F.,** Studies on the morphology of alphaviruses, *Virology,* 95, 501, 1979.
6. **Horzinek, M. C.,** The structure of togaviruses, *Prog. Med. Virol.,* 16, 109, 1973.
7. **Horzinek, M. and Mussgay, M.,** Studies on the nucleocapsid structure of a group A arbovirus, *J. Virol.,* 4, 514, 1969.
8. **Morgan, C., Howe, C., and Rose, H. M.,** Structure and development of viruses as observed in the electron microscope. V. Western equine encephalomyelitis virus, *J. Exp. Med.,* 113, 219, 1961.
9. **Murphy, F. A., Harrison, A. K., Gary, G. W., Jr., Whitfield, S. G., and Forrester, F. T.,** St. Louis encephalitis virus infection of mice. Electron microscopic studies of central nervous system, *Lab. Invest.,* 19, 652, 1968.
10. **Norrby, E.,** Rubella virus, *Virol. Mongr.,* 7, 115, 1969.
11. **Pfefferkorn, E. R. and Shapiro, D.,** Reproduction of togaviruses, in *Comprehensive Virology,* Vol. 2, Fraenkel-Conrat, H. and Wagner, R. R., Eds., Plenum Press, New York, 1974, 171.
12. **Simpson, R. W. and Hauser, R. E.,** Basic structure of group A arbovirus strains Middleburg, Sindbis and Semliki Forest examined by negative staining, *Virology,* 34, 358, 1968.
13. **Smith, T. J., Brandt, W. E., Swanson, J. L., McCown, J. M., and Buescher, E. L.,** Physical and biological properties of dengue-2 virus and associated antigens, *J. Virol.,* 5, 524, 1970.
14. **Strauss, J. H. and Strauss, E. G.,** Togaviruses, in *The Molecular Biology of Animal Viruses,* Vol. 1, Nayak, D. P., Ed., Marcel Dekker, New York, 1977, 111.
15. **von Bonsdorff, C. H. and Harrison, S. C.,** Sindbis virus glycoproteins form a regular icosahedral surface lattice, *J. Virol.,* 16, 141, 1975.

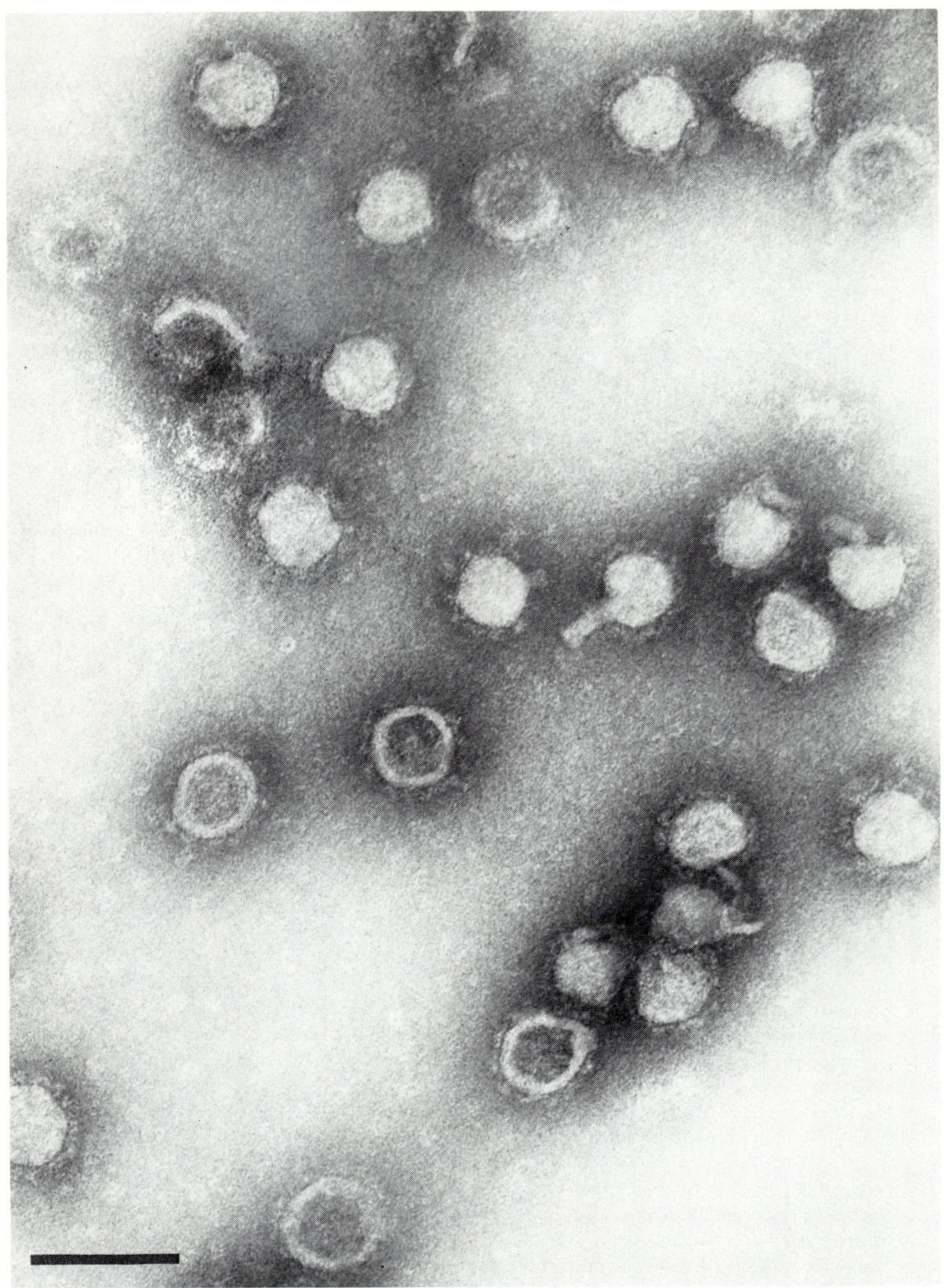

PLATE 1. Morphology of the *Alphavirus*, eastern equine encephalitis virus. Envelope and surface projections are evident. All bars equal 100 nm.

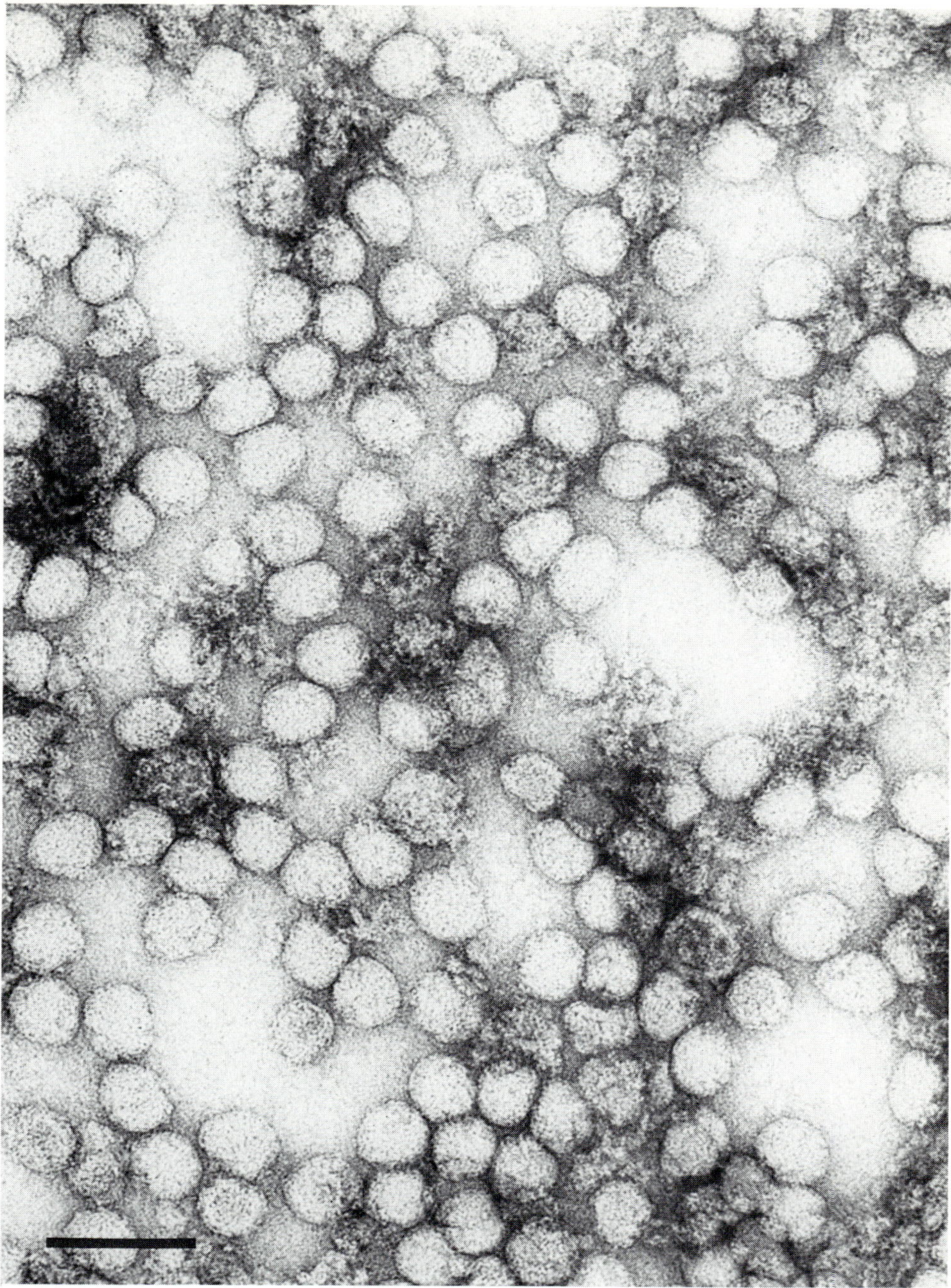

PLATE 2. Morphology of the *Flavivirus*, yellow fever virus. Particles are morphologically indistinct except that they are compact and relatively homogeneous in size.

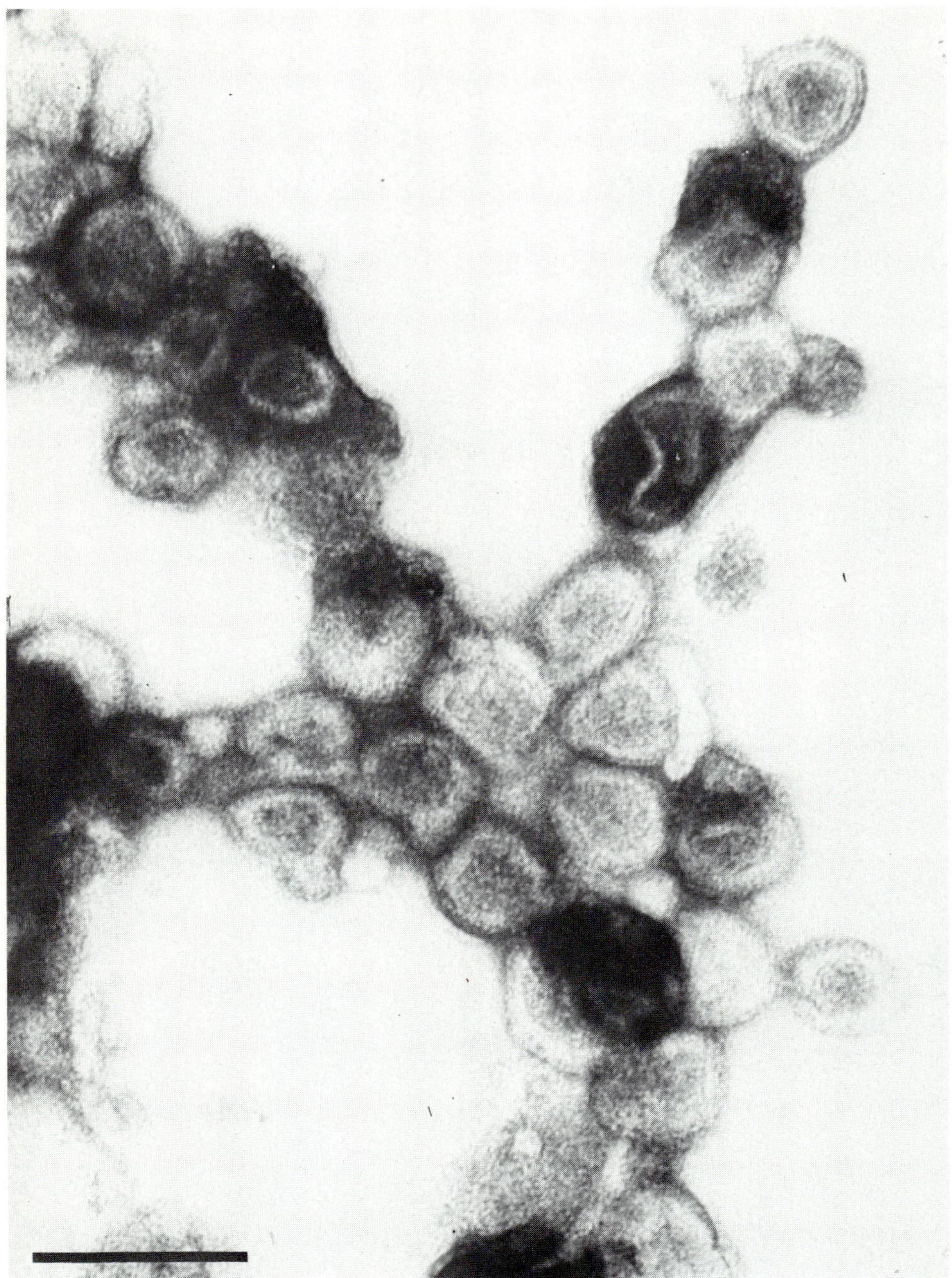

PLATE 3. Rubella virus. Close examination of particles in this group shows that some viruses have a discernable envelope with closely packed surface projections. Viruses of the togaviridae frequently appear nondescript after negative-stain EM.

Part III
RNA Viruses with
Helical Nucleocapsids and
Complex Envelopes

Chapter 8

ORTHOMYXOVIRIDAE

The most important members of the Orthomyxoviridae (Greek orthos = correct, straight; myxa = mucus) family are influenza virus types A and B which have been given the generic status of *Influenza. Influenza* type A has been isolated from humans, swine, horses, and birds, whereas type B has been isolated only from humans. *Influenza* viruses are typed as A or B on the basis of the antigenic specificity of their internal ribonucleoprotein ("soluble antigen") and matrix protein. External proteins (the hemagglutinin and the neuraminidase) comprise additional antigens of *Influenza* type A and B viruses. A number of subgroups have been recognized among hemagglutinin and neuraminidase antigens of *Influenza* A virus, and to a lesser extent, antigenic variation also occurs in the external antigens of *Influenza* B virus. Strains of influenza A or B viruses are identified on the basis of differences in the antigenic subgroups of hemagglutinin and neuraminidase, which are capable of independent variation from each other.

Influenza C virus structurally and chemically resembles *Influenza* A and B viruses in many ways and like *Influenza* B has been isolated from humans. However, it appears to lack a neuraminidase and has a novel receptor specificity which has not yet been chemically identified. Particles also often have an unusual honeycomb-like surface morphology not common to *Influenza* A or B virus. For these reasons *Influenza* C is presently considered to be different enough to be accorded the status of a provisional genus in the Orthomyxoviridae family.

Viruses of freshly isolated strains of *Influenza* are often heterogeneous in size and shape. Filamentous forms several hundred nm in length are common, as are other bizarre forms. They can, nevertheless, be differentiated from other viruses by the presence of evenly spaced spikes covering the entire virus surface. These spikes may be seen projecting end-on toward the viewer or as a "fringe" surrounding an electron-lucent center. They project about 9 nm from the surface of intact particles, are 4 to 8 nm wide, and are spaced 7 to 8 nm apart. The spike layer contains the external influenza antigens: hemagglutinin and a virus-coded enzyme which has been identified in *Influenza* A and B as a neuraminidase. The hemagglutinin attaches to target cells, and agglutinates red blood cells from a variety of animal species. Neuraminidase hydrolyzes neuraminic acid residues from mucoproteins and allows virus to elute from cell receptors. Although influenza C virus also has an enzyme which destroys the influenza receptor, it does not appear to be a neuraminidase.

Beneath the spike layer is a characteristic lipid-containing envelope enclosing a matrix protein and a helical nucleocapsid formed of protein and single-stranded RNA. A viral RNA-dependent RNA polymerase activity is associated with the nucleocapsid of *Influenza* A virus, and probably with the nucleocapsids of types B and C as well. The envelope is 6 to 10 nm thick. When this layer is disrupted or penetrated by stain, a nucleocapsid can sometimes be seen as folded parallel bands. Free-lying nucleocapsids can be obtained by treating virions with detergent. They appear as small supercoiled helices of varying size. There are 8 RNP segments, each of which codes for a separate protein.

Influenza C virus may appear to be identical in morphology to *Influenza* A or B, or may be stained so that a hexagonal structure of the virus surface is clearly evident. Usually both types of particles are seen in the same field.

REFERENCES

1. **Choppin, P. W.**, Multiplication of two kinds of influenza A_2 virus particles in monkey kidney cells, *Virology*, 21, 342, 1963.
2. **Choppin, P. W. and Compans, R. W.**, The structure of influenza virus, in *The Influenza Viruses and Influenza*, Kilbourne, E. D., Ed., Academic Press, New York, 1975, 15.
3. **Choppin, P. W., Murphy, J. S., and Stoeckenius, W.**, The surface structure of influenza virus filaments, *Virology*, 13, 548, 1961.
4. **Compans, R. W. and Choppin, P. W.**, Orthomyxoviruses and paramyxoviruses, in *Ultrastructure of Animal Viruses and Bacteriophages: An Atlas*, Dalton, J. A. and Haguenau, F., Eds., Academic Press, New York, 1972, 213.
5. **Compans, R. W. and Choppin, P. W.**, Reproduction of myxoviruses, in *Comprehensive Virology*, Vol. 4, Fraenkel-Conrat, H. and Wagner, R. R., Eds., Plenum Press, New York, 1975, 179.
6. **Dowdle, W. R., Davenport, F. M., Fukumi, H., Schild, G. C., Tumova, B., Webster, R. G., and Zakstelskaja, L.**, Orthomyxoviridae, *Intervirology*, 5, 245, 1975.
7. **Duesberg, P. H.**, Distinct subunits of the ribonucleoprotein of influenza virus, *J. Mol. Biol.*, 42, 485, 1969.
8. **Horne, R. W., Waterson, A. P., Wildy, P., and Farnham, A. E.**, The structure and composition of the myxoviruses. I. Electron microscope studies of the structure of the myxovirus particles by negative staining techniques, *Virology*, 11, 79, 1960.
9. **Hoyle, L.**, The influenza viruses, *Virol. Monogr.*, 4, 1968.
10. **Hoyle, L., Horne, R. W., and Waterson, A. P.**, The structure and composition of the myxoviruses. II. Components released from the influenza virus particle by ether, *Virology*, 13, 448, 1961.
11. **Kendal, A. P.**, A comparison of "influenza C" with prototype myxoviruses: receptor-destroying activity (neuraminidase) and structural polypeptides, *Virology*, 65, 87, 1975.
12. **Laver, W. G.**, Structural studies on the protein subunits from three strains of influenza virus, *J. Mol. Biol.*, 9, 109, 1964.
13. **Laver, W. G. and Valentine, R. C.**, Morphology of the isolated hemagglutinin and neuraminidase subunits of influenza virus, *Virology*, 38, 105, 1969.
14. **Martin, M. L., Palmer, E. L., and Kendal, A. P.**, Lack of characteristic hexagonal surface structure on a newly isolated influenza C virus, *J. Clin. Micro.*, 6, 84, 1977.
15. **Pons, M. W., Schulze, I. T., and Hirst, G. K.**, Isolation and characterization of the ribonucleoprotein of influenza virus, *Virology*, 39, 250, 1969.
16. **Schulze, I. T.**, Structure of the influenza virion, *Adv. Virus Res.*, 18, 1, 1973.
17. **Schulze, I. T., Pons, M. W., and Hirst, G. K.**, The RNA and proteins of influenza virus, in *The Biology of Large RNA Viruses*, Barry, R. D. and Mahy, B. W. J., Eds., Academic Press, New York, 1970, 324.
18. **Tiffany, J. M. and Blough, H. A.**, Models of structure of the envelope of influenza virus, *Proc. Natl. Acad. Sci. U.S.A.*, 65, 1105, 1970.
19. **Webster, R. G. and Laver, W. G.**, Antigenic variation in influenza virus: biology and chemistry, *Prog. Med. Virol.*, 13, 271, 1971.
20. **WHO Study Group**, A revised system of nomenclature for influenza viruses, *Bull. W.H.O.*, 45, 119, 1971.
21. **Wrigley, N. G., Skehel, J. J., Charlwood, P. A., and Brand, C. M.**, The size and shape of influenza virus neuraminidase, *Virology*, 51, 525, 1973.

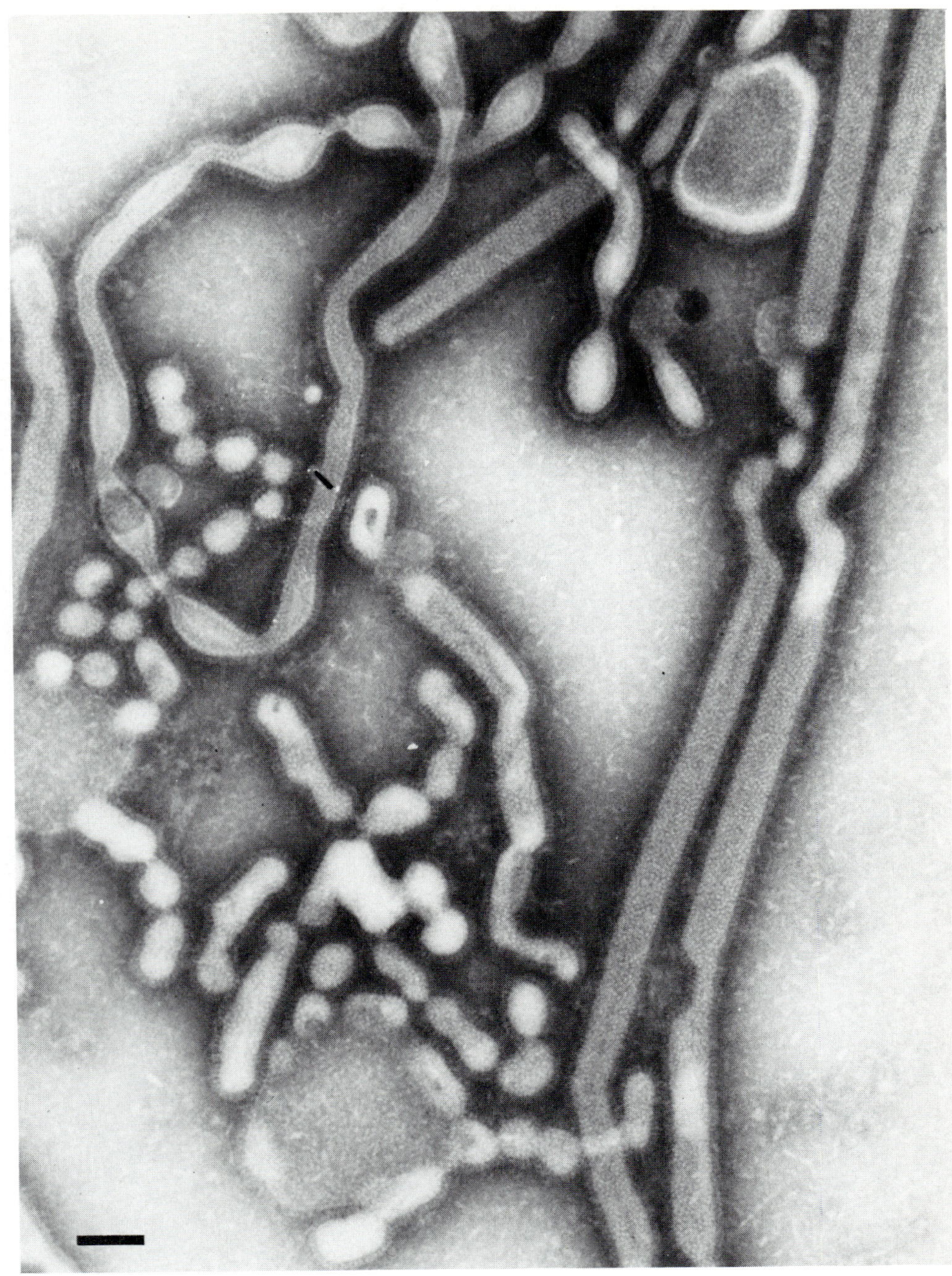

PLATE 1. Filamentous forms of *Influenza* A. These forms are characteristic of virions as they appear after first passage in chicken amniotic fluid. All bars equal 100 nm.

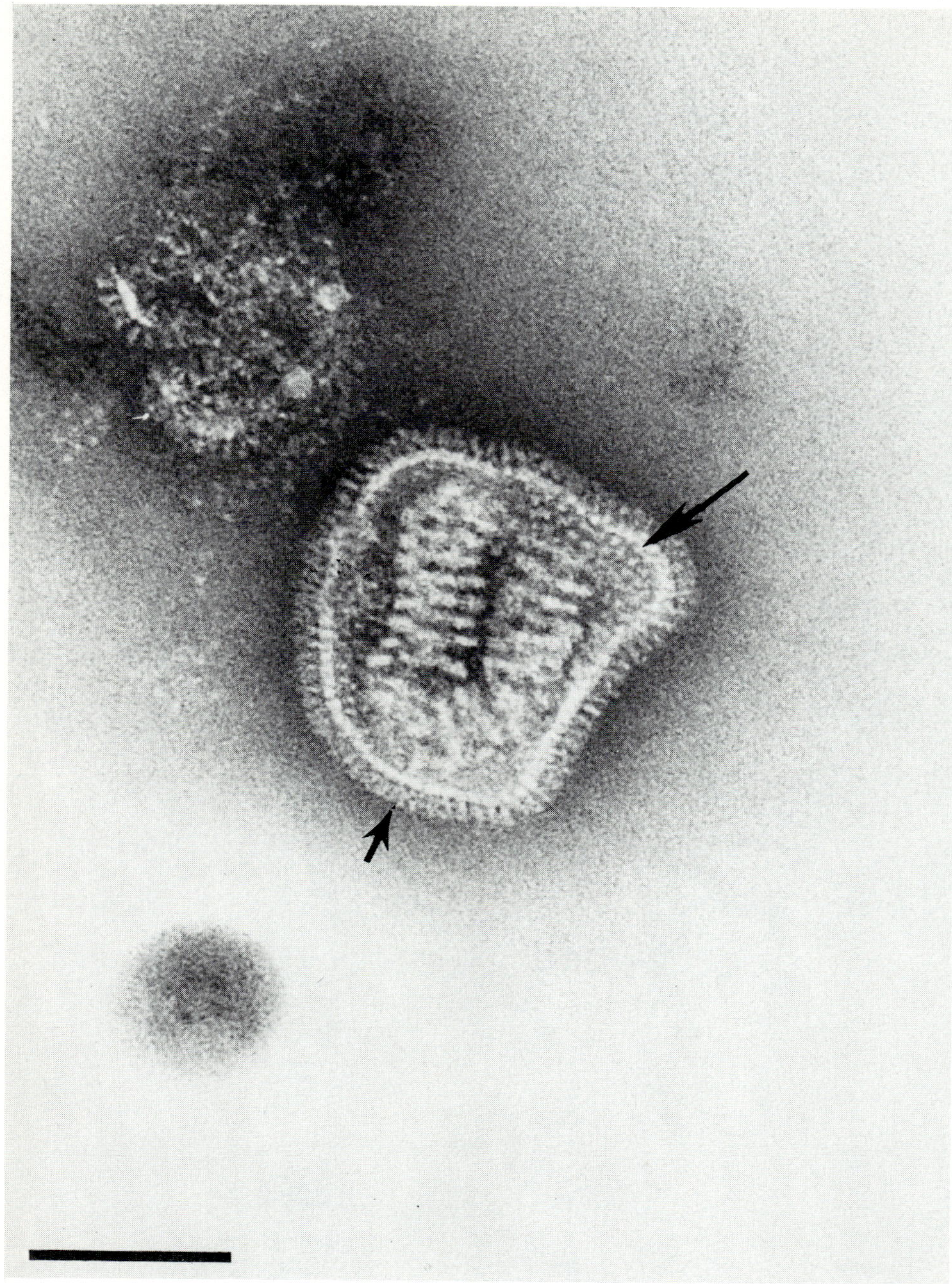

PLATE 2. *Influenza* A particle. The envelope (long arrow), spikes (short arrow) and internal ribonucleo-protein can easily be discerned.

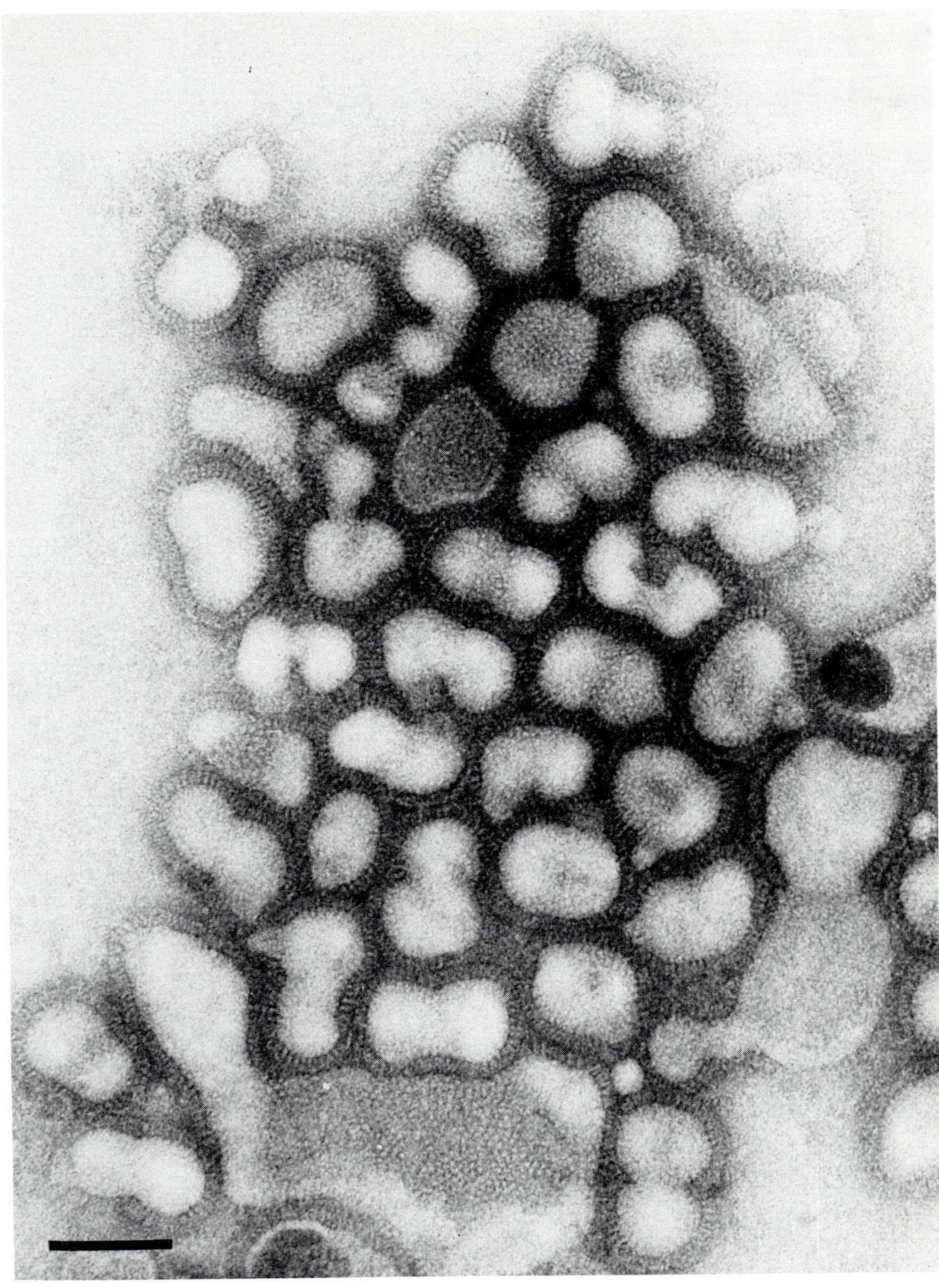

PLATE 3. *Influenza* A virions as they appear after several passages in eggs. Particles tend to be more uniform in size (about 100 nm), and filamentous forms are not prevalent.

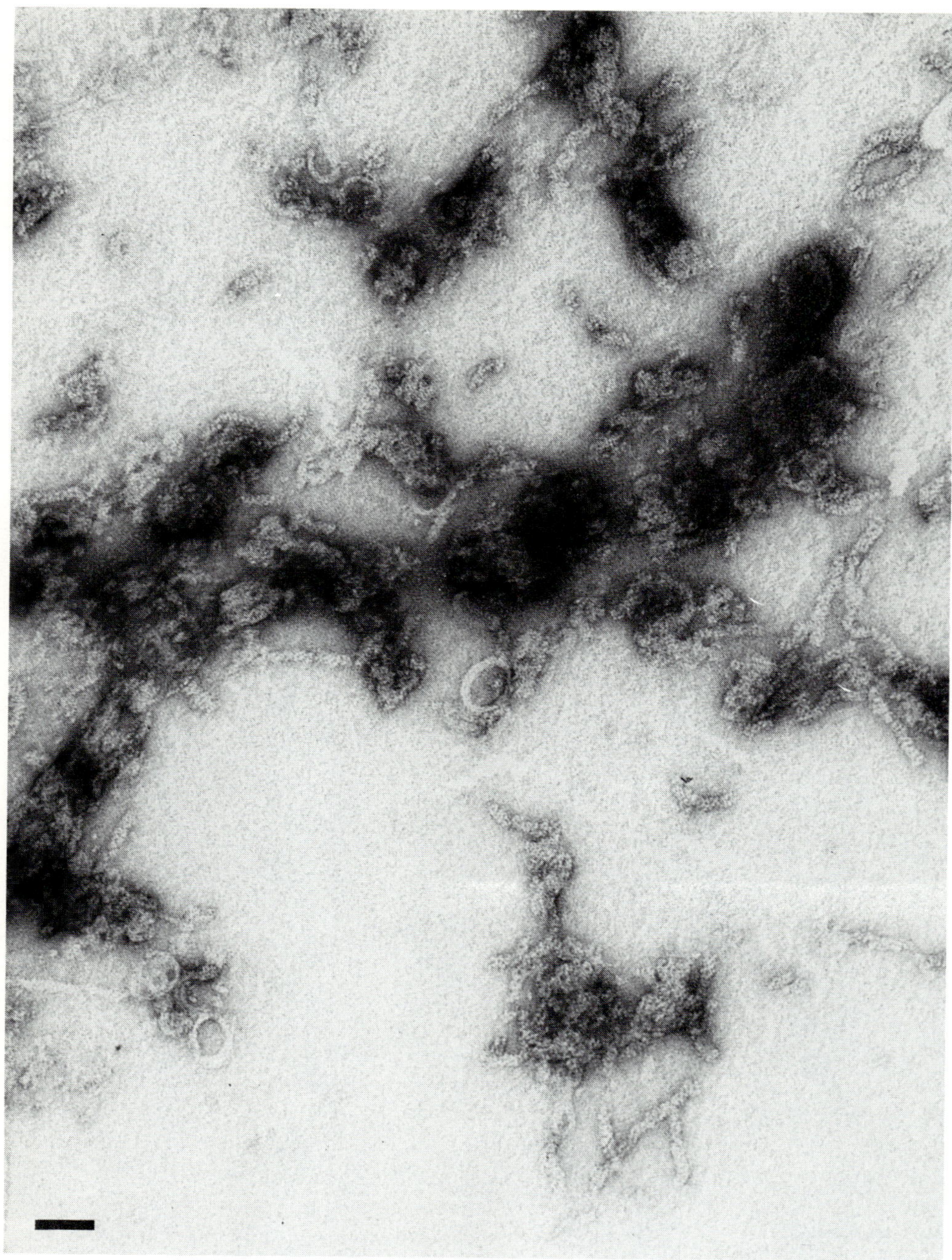

PLATE 4. Free-lying *Influenza* A ribonucleoprotein separated from detergent-treated virions. Virions contain 8 RNP segments; each RNA codes for separate viral proteins.

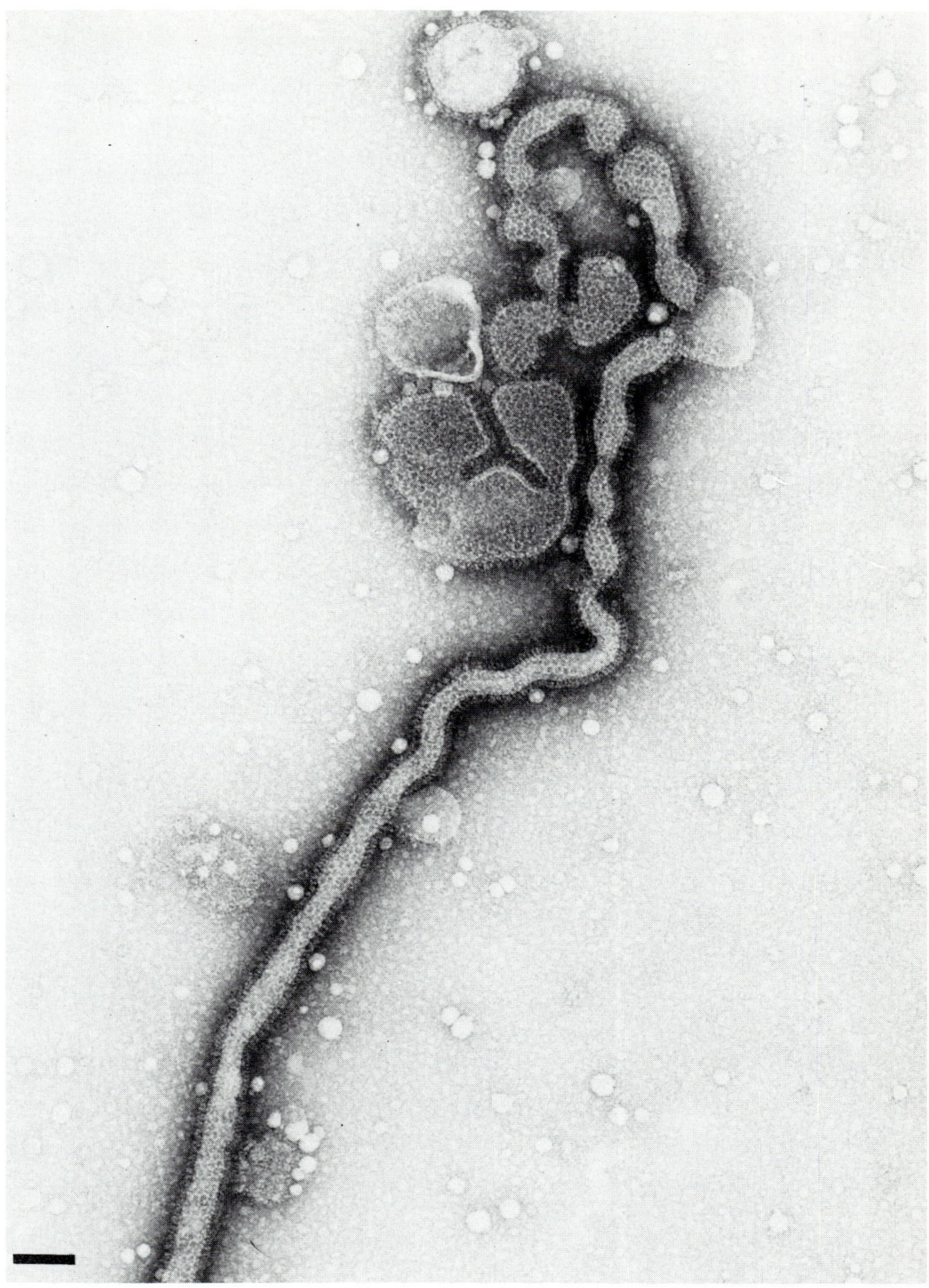

PLATE 5. Influenza C. When initially isolated, these particles closely resemble *Influenza* A and B, but after several passages in eggs a distinctive hexagonal surface structure becomes evident.

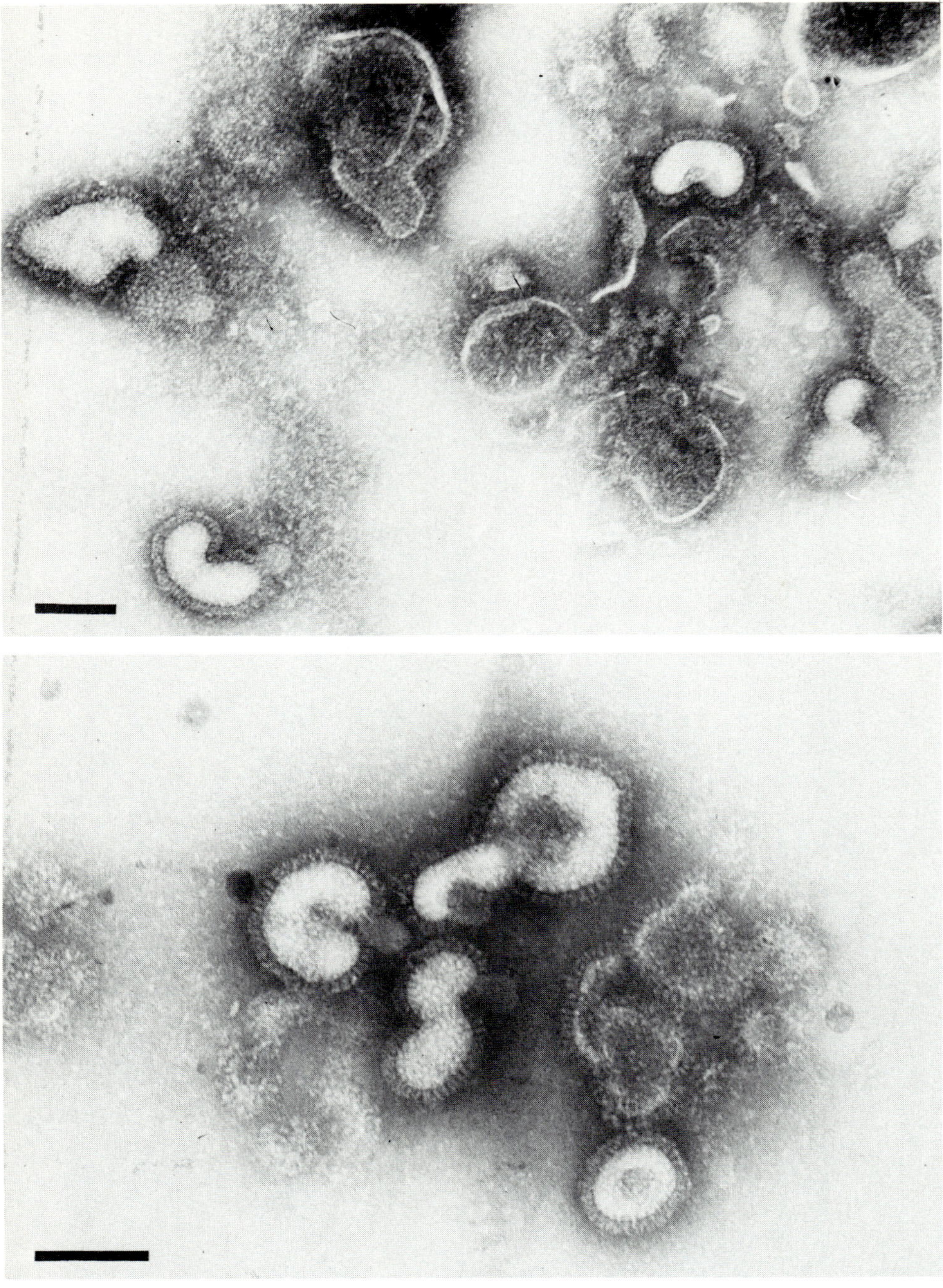

PLATE 6. Influenza from chicken egg amniotic fluid. Amniotic fluid has little cell debris and viruses present in the fluid are easily seen. Small proteins in the fluid also help spread the negative stain to make it an ideal preparation for EM examination.

Chapter 9

PARAMYXOVIRIDAE

The Paramyxoviridae family consists of three genera important to humans: *Paramyxovirus,* comprising parainfluenza virus types 1 through 4 (type 4 has subtypes A and B), the Newcastle disease virus group, and mumps virus; *Morbillivirus,* comprising the measles (rubeola) virus group; and *Pneumovirus,* which includes respiratory syncytial virus and related agents. *Paramyxovirus* and *Morbillivirus* are identical in morphology *Pneumovirus,* although having some morphological features in common with other Paramyxoviridae, differs enough to be distinguishable from them by EM.

The *Paramyxovirus* (Greek para, "by the side of") genus encompasses a number of viruses which cause clinical disease in humans. Significant serological cross-reactivity occurs among viruses within the genus, but there is no detectable cross-reactivity with *Morbillivirus, Pneumovirus* or other members of the family Orthomyxoviridae. All *Paramyxovirus* species have three type-specific antigens: a hemagglutinin surface antigen, a neuraminidase surface antigen, and an internal nucleocapsid antigen. However, for mumps virus the hemagglutinin is also commonly called the V (viral) antigen and the ribonucleoprotein antigen is referred to as the S (soluble) antigen.

The *Paramyxovirus* and *Morbillivirus* members are quasi-spherical and average about 125 to 250 nm in diameter; however, some individual particles may be very small (100 nm) or very large (approaching 1 μm). The nucleocapsid is surrounded by an envelope about 10 nm thick. The envelope is covered with clearly seen regularly arrayed spikes consisting of the hemagglutinin and neuraminidase antigens. The nucleocapsid, unlike that of the Orthomyxoviruses, is a continuous helix. It has a width of approximately 18 nm and a central hole of about 4 nm; a periodicity of 4 nm has been determined for the serrations. This structure is quite distinctive when seen by negative-contrast electron microscopy, resembling a herringbone pattern, and provides a means of distinguishing *Paramyxovirus* and *Morbillivirus* from *Influenza* A or B viruses. It is also important to note that these viruses seldom exhibit the filamentous forms characteristic of *Influenza.*

The *Morbillivirus* (Morbus = disease) measles and related viruses are identical in morphology to members of the *Paramyxovirus* genus when viewed by negative-contrast electron microscopy. However, they do not have antigens in common with the Paramyxoviruses and have several distinct features. Measles virus possesses a hemagglutinin corresponding to its surface projections, but the projections do not have neuraminidase activity. Further, measles virus contains a lipid soluble factor which hemolyzes monkey erythrocytes and induces giant-cell formation of cultured cells. Antigenic variations among measles virus strains have not been detected.

Respiratory syncytical virus (RSV), the type species of the *Pneumovirus* genus (pneumo = breath), is one of the most important pathogens which cause disease in young children. It affects the entire respiratory tract, producing a variety of syndromes such as the common cold, laryngotracheobronchitis (croup) and a host of other maladies. It was named RSV because it characteristically produces large syncytial masses in tissue culture. RSV has a number of properties that separate it from *Paramyxovirus* and measles virus. It possesses surface projections which are club-like in appearance and are regularly spread, but it lacks hemagglutinin, neuraminidase, or hemolytic activity. RSV readily forms filaments and is often pleomorphic. The particles are roughly spherical and average 90 to 130 nm in diameter; very large or very small particles are rarely seen. The nucleocapsid is about 14 nm wide, with a periodicity of serrations of about 7 nm.

REFERENCES

Paramyxoviridae

1. **Bloth, B., Espmark, A., Norrby, E., and Gard, S.,** The ultrastructure of respiratory syncytial (RS) virus, *Arch. Gesamte Virusforsch.,* 13, 582, 1963.
2. **Choppin, P. W. and Compans, R. W.,** Reproduction of paramyxoviruses, in *Comprehensive Virology,* Vol. 4, Fraenkel-Conrat, H. and Wagner, R. R., Eds., Plenum Press, New York, 1975, 95.
3. **Choppin, P. W. and Stoeckenius, W.,** The morphology of SV5 virus, *Virology,* 23, 195, 1964.
4. **Compans, R. W. and Choppin, P. W.,** The structure and assembly of influenza and parainfluenza viruses, in *Comparative Virology,* Maramorosch, K. and Kurstak, E., Eds., Academic Press, New York, 1971, 407.
5. **Compans, R. W., Holmes, K. V., Dales, S., and Choppin, P. W.,** An electron microscopic study of moderate and virulent virus-cell interactions of the parainfluenza virus SV5, *Virology,* 30, 411, 1966.
6. **Finch, J. T. and Gibbs, A. J.,** Observations on the structure of the nucleocapsids of some paramyxoviruses, *J. Gen. Virol.,* 6, 141, 1970.
7. **Fraser, K. B. and Martin, S. J.,** *Measles Virus and its Biology,* Academic Press, New York, 1978.
8. **Horne, R. W. and Waterson, A. P.,** A helical structure in mumps, Newcastle disease and Sendai viruses, *J. Mol. Biol.,* 2, 75, 1960.
9. **Hosaka, Y., Kitamo, H., and Ikeguchi, S.,** Studies on the pleomorphism of HVJ virions, *Virology,* 29, 205, 1966.
10. **Kingsbury, D. W.,** Paramyxovirus replication, *Curr. Top. Microbiol. Immunol.,* 59, 1, 1972.
11. **Kingsbury, D. W.,** Paramyxoviruses, in *The Molecular Biology of Animal Viruses,* Nayak, D. P., Ed., Marcel Dekker, New York, 1977, 349.
12. **Klenk, H. D. and Choppin, P. W.,** Chemical composition of the parainfluenza virus SV5, *Virology,* 37, 155, 1969.
13. **Kolakofsky, D., Boy de la Tour, E., and Bruschi, A.,** Self-annealing of Sendai virus RNA, *J. Virol.,* 14, 33, 1974.
14. **Nakai, T., Shand, F. L., and Howatson, A. F.,** Development of measles virus *in vitro, Virology,* 38, 50, 1969.
15. **Norrby, E. C. J. and Magnusson, P.,** Some morphological characteristics of the internal component of measles virus, *Arch. Gesamte Virusforsch.,* 17, 443, 1965.
16. **Vorkunova, G. K., Pashova, V. A., Klimenko, S. M., Gushchin, B. V., and Bukrinskaya, A. G.,** The properties of intracellular paramyxovirus ribonucleoprotein in a nonpermissive system, *Arch. Gesamte Virusforsch.,* 46, 44, 1974.
17. **Waterson, A. P.,** Two kinds of myxovirus, *Nature (London),* 193, 1163, 1962.
18. **Zakstelskaya, L. Y., Almeida, J. D., and Bradstreet, C. M. P.,** The morphological characterization of respiratory syncytial virus by a simple electron microscopy technique, *Acta Virol.,* 11, 420, 1967.

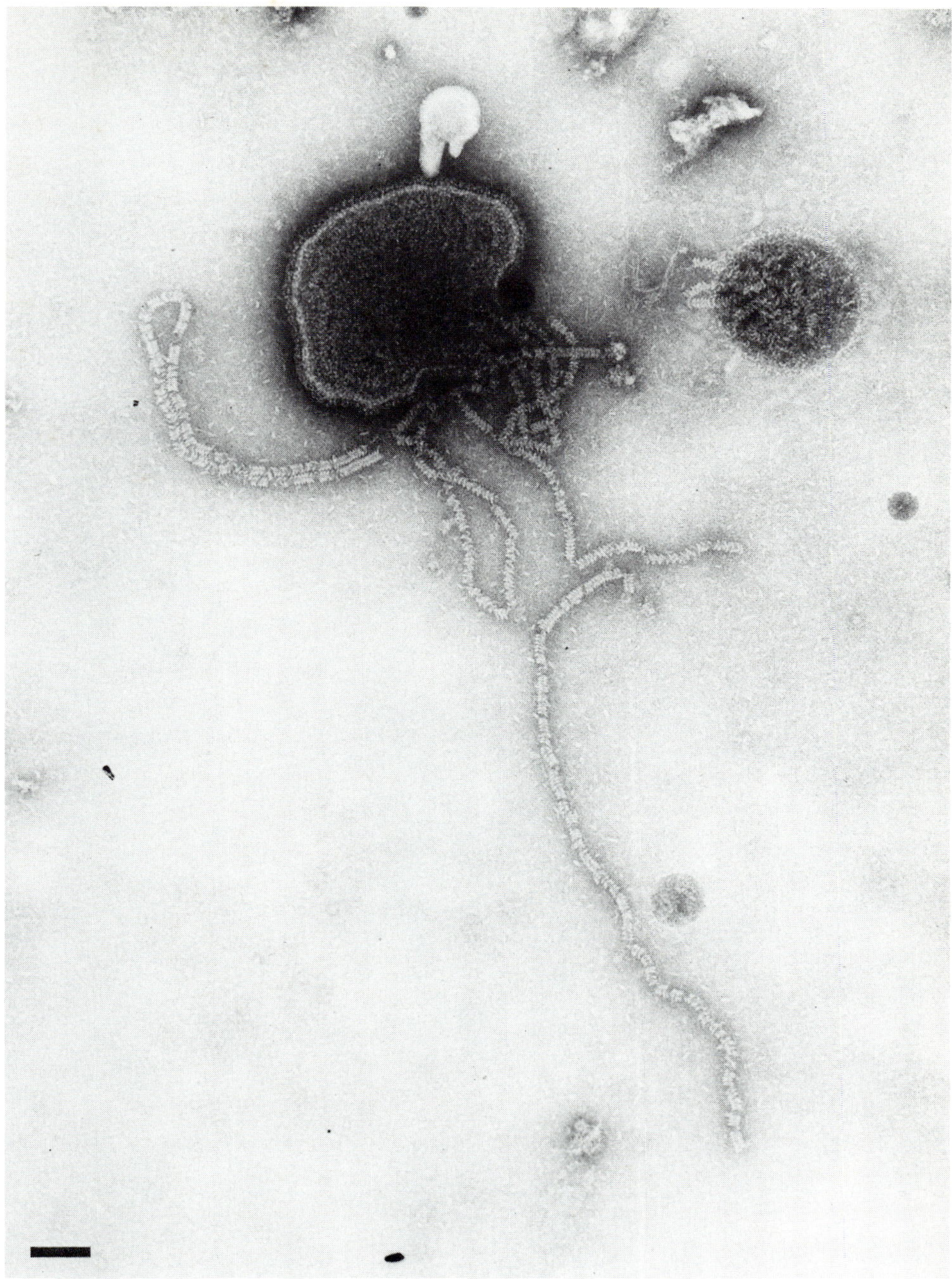

PLATE 1.   Parainfluenza virus. Disrupting particle is releasing helical nucleocapsid with morphology resembling a herringbone. Free-lying nucleocapsid helices are also seen in the same field. The disrupting particle also has an envelope from which surface projections radiate. All bars equal 100 nm.

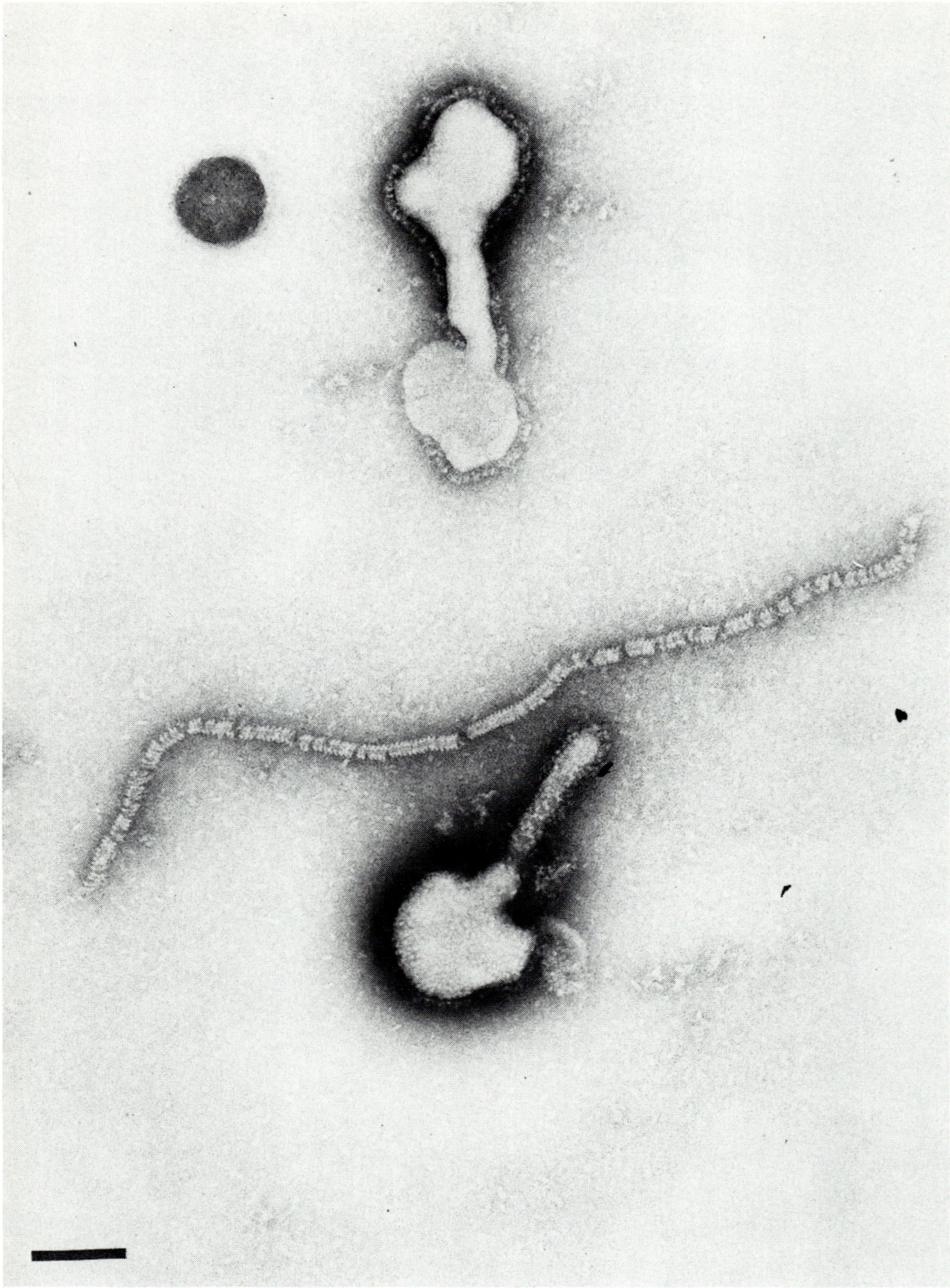

PLATE 2.   Parainfluenza virus. Two intact particles and free-lying nucleocapsid.

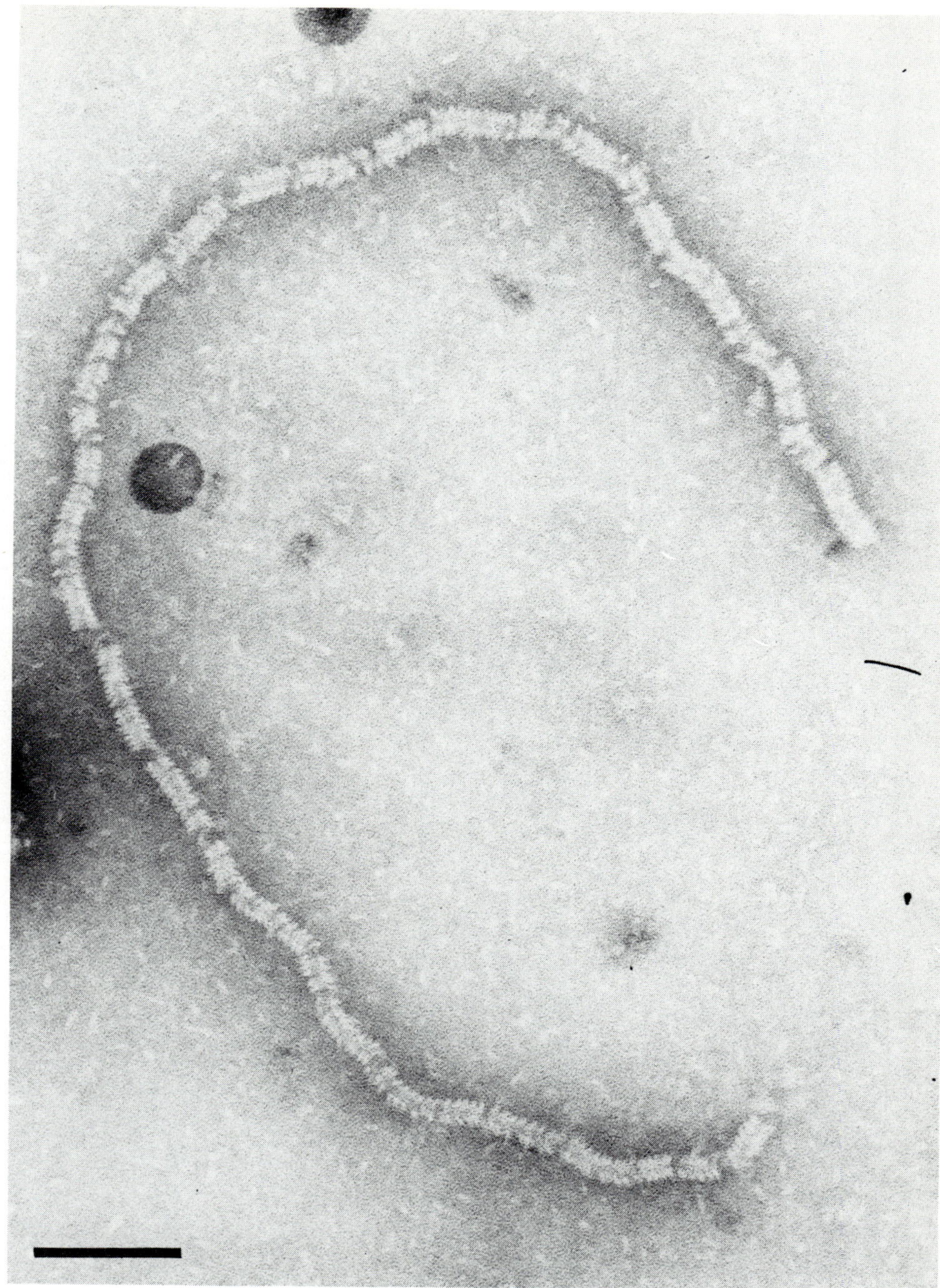

PLATE 3.   High magnification of a free lying helical paramyxovirus nucleocapsid. The helix is continuous with a width of about 18 nm. The periodicity of serrations is about 4 nm as is the central hole.

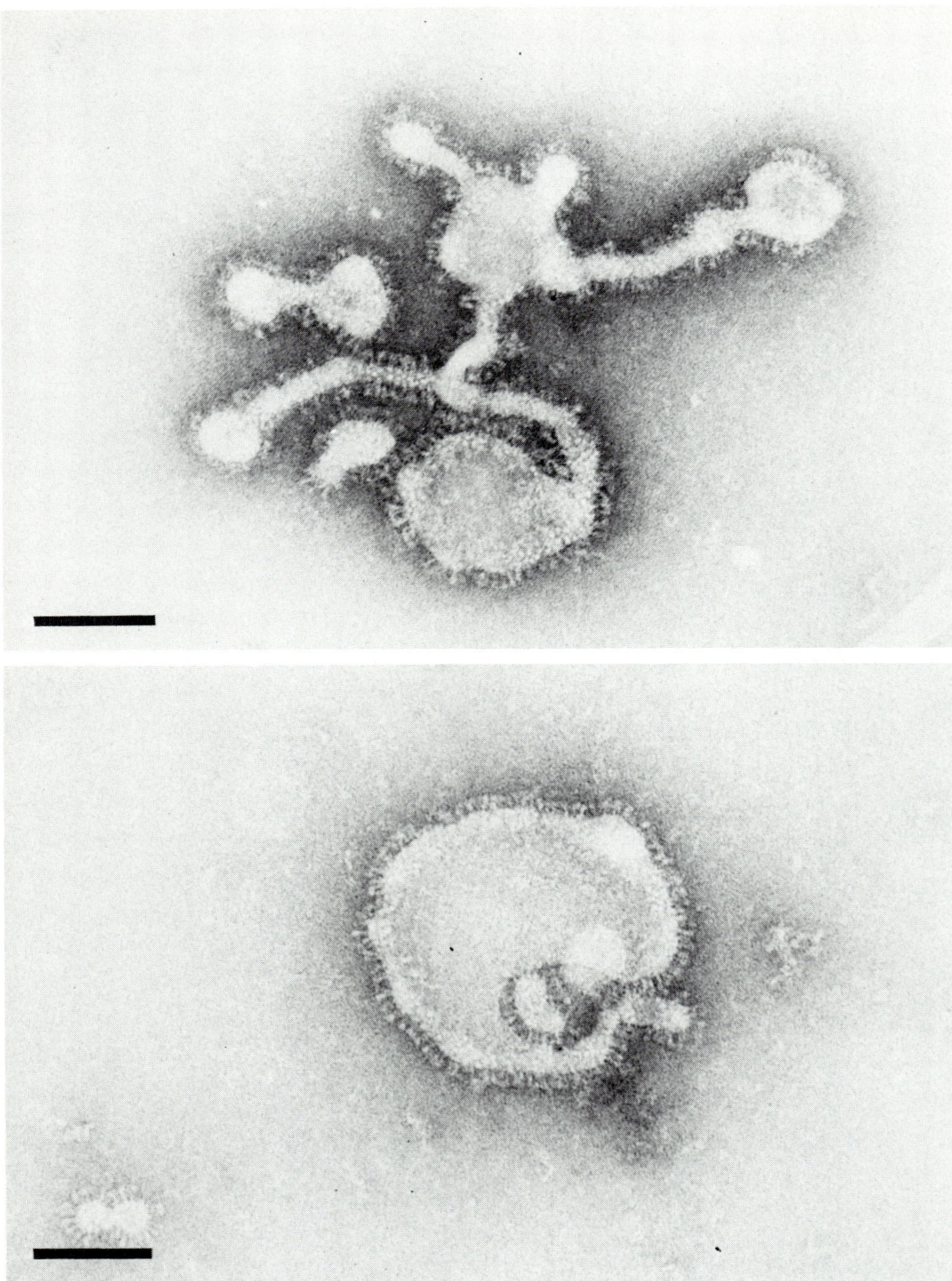

PLATE 4.   Composite showing two forms of respiratory syncytial virus. The virus is pleomorphic and is difficult to discern in tissue culture preparations. It is frequently confused with cell membrane fragments. The virus envelope is not easily seen, but surface projections are long and evenly spaced on many particles.

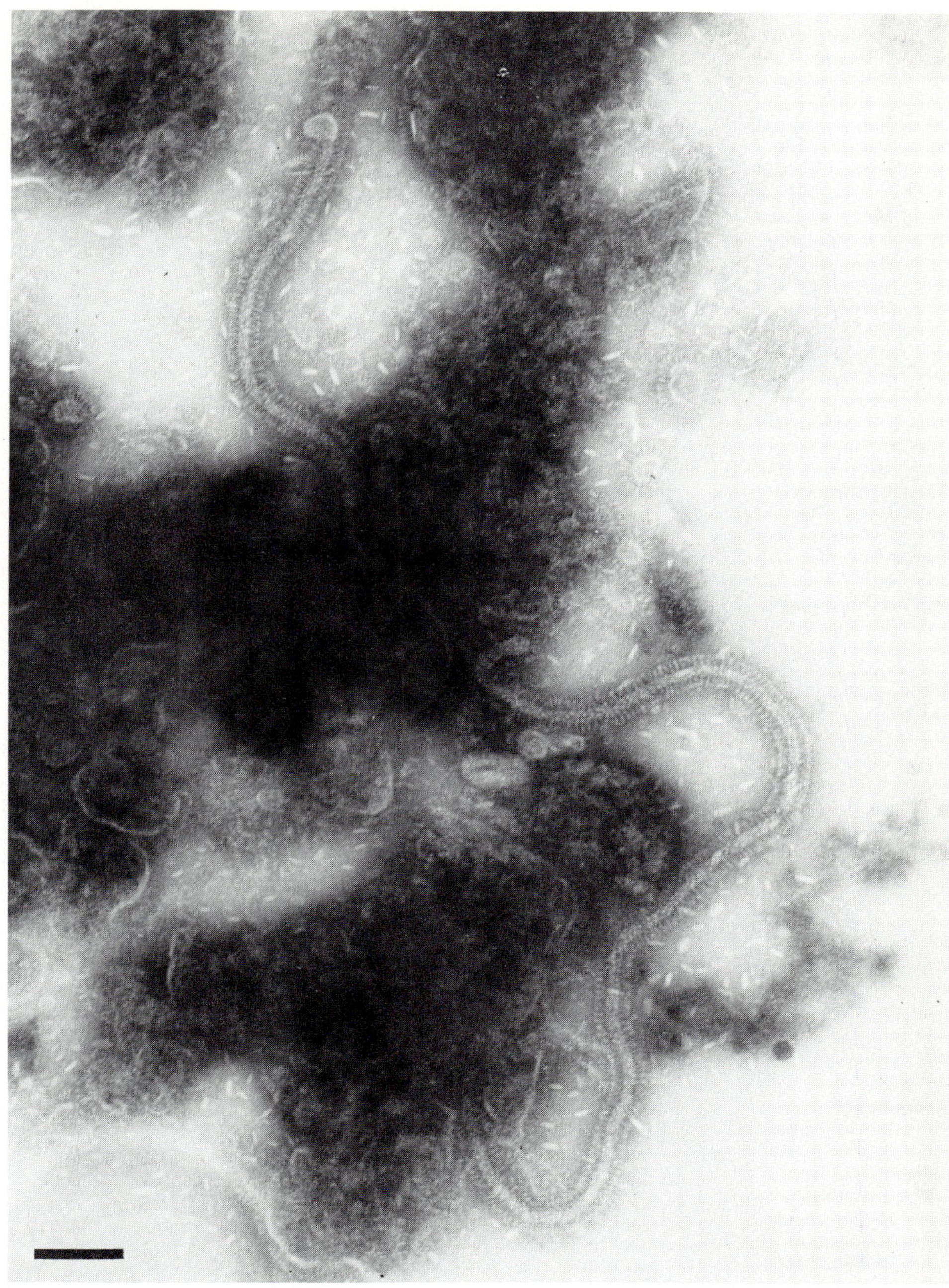

PLATE 5.   Respiratory syncytial virus. Isolate from a throat washing. The long filamentous form is seen among cell debris.

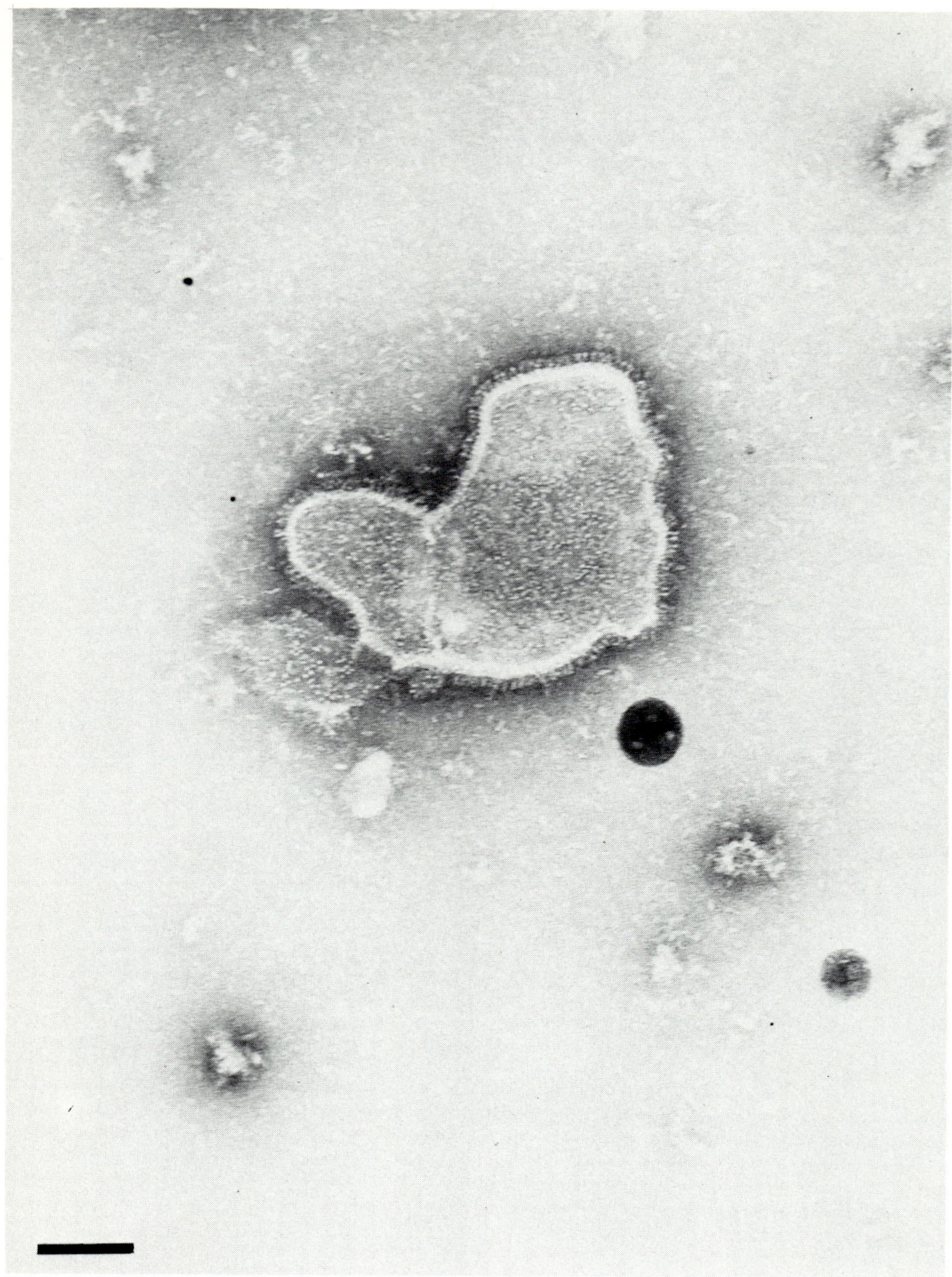

PLATE 6.   Pleomorphic form of RSV. Surface projections can be seen end on.

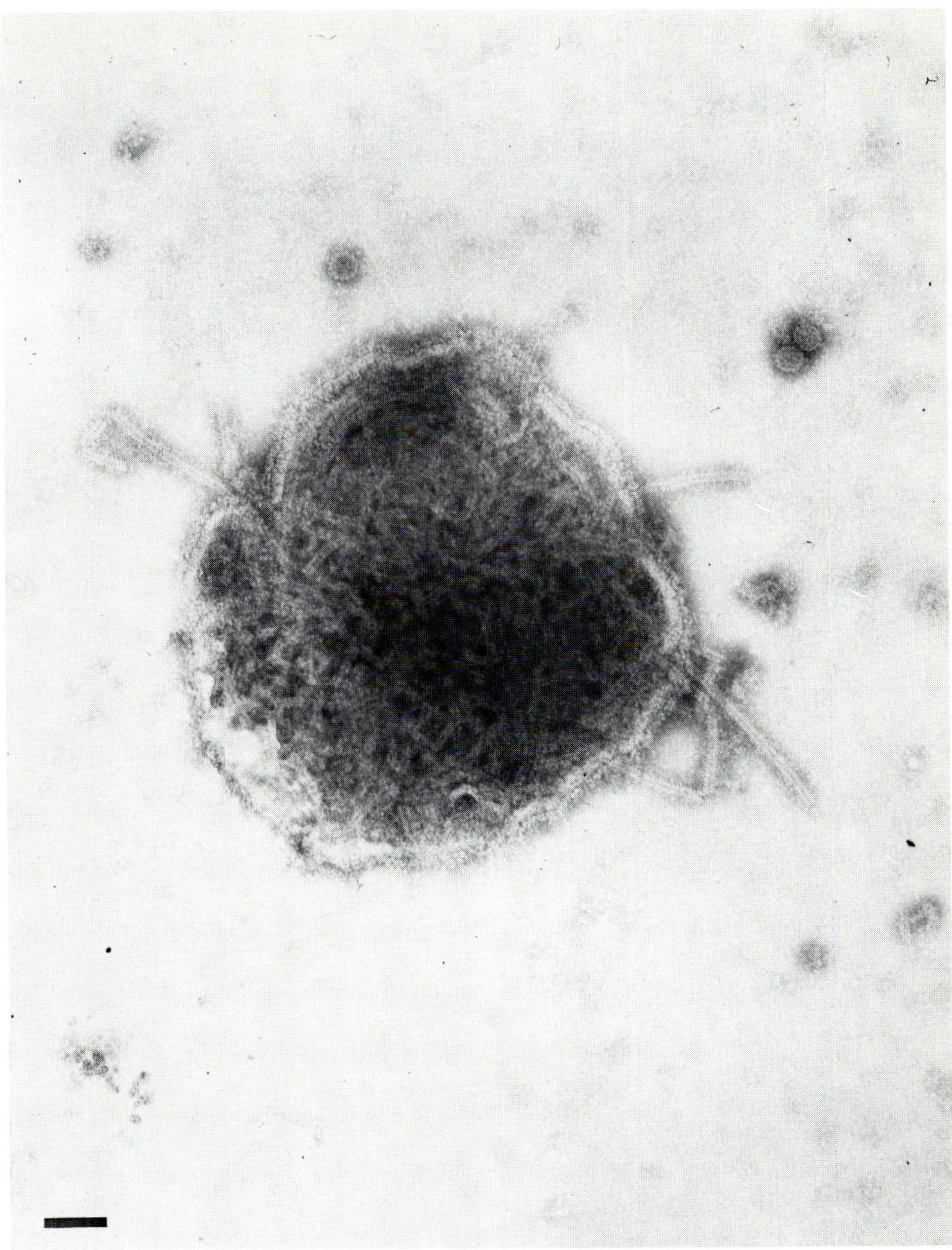

PLATE 7.   High magnification of mumps virus from monkey kidney cells which are also latently infected with SV40 (a papovavirus).

Chapter 10

RETROVIRIDAE

The Retroviridae family consists of a large number of RNA tumor viruses and related viruses which have a virion-associated RNA-dependent DNA polymerase. Retro, Latin meaning backwards, refers to this endogenous reverse transcriptase.

Viruses of the Retroviridae are about 100 nm in diameter with a lipoprotein envelope which has surface projections. The envelope encloses an icosahedral core which, in turn, encloses a helical ribonucleoprotein. When seen by negative-stain EM, many of the retroviruses appear pleomorphic, especially the ones causing leukemias and sarcomas in various animal species. Virions appear to be sensitive to changes in osmotic pressure brought about by specimen preparation, and they often are fragile after purification. Many RNA tumor viruses of the subfamily Oncovirinae have the unusual feature of replicating to high titer in various kinds of cells without alteration of the cell. The most common of these, usually seen in thin sections, are C-type particles. Most mammalian species have an associated C-type particle.

In thin section C-type particles have an appearance similar to a target bull's eye with an electron-dense center because the core shell and nucleocapsid are arranged concentrically. Type B oncoviruses, such as mouse mammary tumor virus, differ from C-type particles in that the core of the virion is located eccentrically; it is separated from the outer coat by an electron-lucent space. The type B oncoviruses appear to be more stable to negative-stain technique than are C-type particles.

The general features of the Retroviruses are seen in Plate 1, depicting a purified preparation of the oncovirus, feline leukemia virus.

REFERENCES

1. **Bernhard, W.,** Electron microscopy of tumor cells and tumor viruses. A review, *Cancer Res.,* 18, 491, 1958.
2. **Bernhard, W.,** The detection and study of tumor viruses with the electron microscope, *Cancer Res.,* 20, 712, 1960.
3. **Bonar, R. A., Heine, U., and Beard, J. W.,** Virus of avian myeloblastosis (BAI strain A). XXIII. Morphology of virus and comparison with strain R (erythroblastosis), *Natl. Cancer Inst. Monogr.,* 17, 589, 1964.
4. **Bonar, R. A., Heine, U., Beard, D., and Beard, J. W.,** Structure of BAI strain A (myeloblastosis) avian tumor virus, *J. Natl. Cancer Inst.,* 30, 949, 1963.
5. **Chopra, H. C. and Mason, M. M.,** A new virus in a spontaneous mammary tumor of a rhesus monkey, *Cancer Res.,* 30, 2081, 1970.
6. **Dalton, A. J., Potter, M., and Mervin, R. M.,** Some ultrastructural characteristics of a series of primary and transplanted plasma-cell tumors of the mouse, *J. Natl. Cancer Inst.,* 26, 1221, 1961.
7. **de Harven, E.,** Morphology of murine leukemia viruses, in *Experimental Leukemia,* Rich, M., Ed., Appleton, New York, 1968, 97.
8. **Gilden, R. V.,** Biology of RNA tumor viruses, in *The Molecular Biology of Animal Viruses,* Vol. 1, Nayak, D. P., Ed., Marcel Dekker, New York, 1977, 435.
9. **Nermut, M. V., Herman, F., and Schafer, W.,** Properties of mouse leukemia viruses. III. Electron microscopical appearance as revealed after conventional preparation techniques as well as freeze drying and freeze etching, *Virology,* 49, 345, 1972.
10. **Rao, P. R., Bonar, R. A., and Beard, J. W.,** Lipids of the BAI strain A avian tumor virus and of the myeloblast host cell, *Exp. Mol. Pathol.,* 5, 374, 1966.
11. **Stromberg, K.,** Surface-active agents for isolation of the core component of avian myeloblastosis virus, *J. Virol.,* 9, 684, 1972.
12. **Tooze, J.,** RNA tumour viruses: morphology, composition and classification, in *The Molecular Biology of Tumor Viruses,* Cold Spring Harbor Laboratory, New York, 1973, 502.

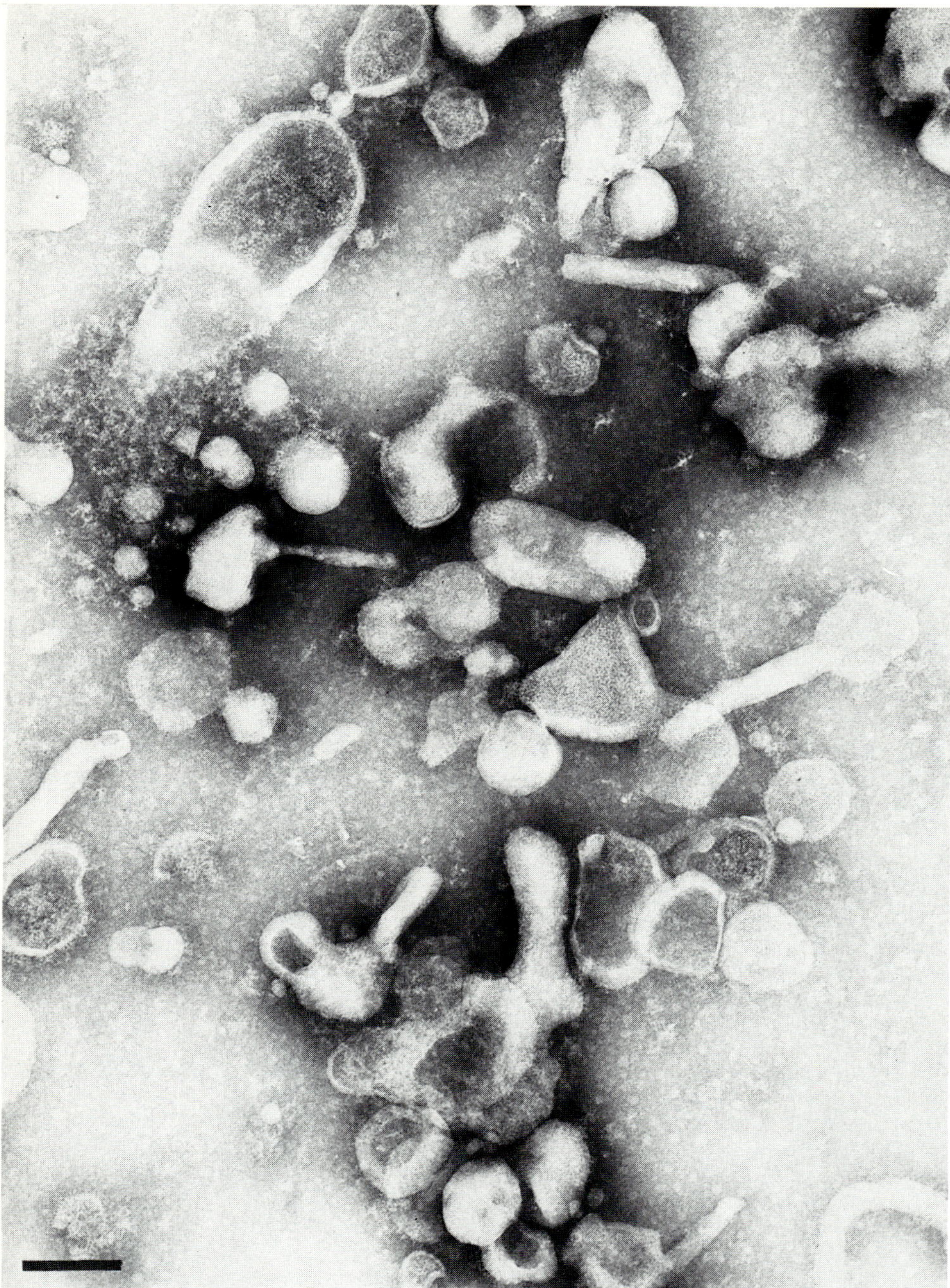

PLATE 1. Survey EM of Feline Leukemia Virus. Particles generally appear pleomorphic and fragile by negative-stain procedures. Bar equals 100 nm.

Chapter 11

RHABDOVIRIDAE AND MARBURG-EBOLA VIRUSES

The family Rhabdoviridae (rhabdo = rod) comprises all known bullet-shaped viruses. These are vesicular stomatitis virus (VSV), rabies virus, and other bullet-shaped viruses isolated from various vertebrates, plants, and insects. Rabies virus infects all warm-blooded mammals and causes a generally fatal disease. It is possible that Marburg and Ebola viruses also belong in this family. Marburg virus and the morphologically similar but antigenically distinct Ebola virus cause hemorrhagic fevers in man which are frequently fatal. These latter two agents are RNA viruses indigenous to Africa which were first isolated in 1967 and 1976, respectively.

The main structural features of VSV (genus *Vesiculovirus)* and rabies (genus *Lyssavirus)* are the bullet or bacilliform shape, an envelope, and a helical nucleocapsid. VSV measures about 170 nm long and 70 nm wide. The nucleoprotein is in the form of a single helix of about 30 turns and capped by 4 or 5 turns of diminishing size at the bullet-shaped end. The nucleocapsid is tightly packed into an envelope which has evenly spaced surface projections. Unfractionated preparations also contain T (truncated) particles which are about half the length of B (complete or bullet) particles. The T particles are not infectious, but they interfere with the infectivity of B particles. Rabies virus is more bacilliform than VSV, is generally longer (200 to 300 nm), and varies more in diameter (50 to 80 nm). Viruses of the two genera are antigenically distinct.

The Marburg agent, which first caused serious illness in a group of laboratory workers in Marburg, Germany in 1967, is a large rod-like virus frequently seen with one end curled in the form of a 6 or a 9. It was introduced into the laboratory by imported African green monkeys from which the workers were preparing primary tissue culture. Stain-penetrated particles reveal what appears to be ribonucleoprotein surrounded by an envelope which is covered with surface projections. Ebola virus was the cause of an outbreak of hemorrhagic fever in the Sudan and Zaire in 1976. It was found to be serologically distinct from the Marburg agent, although both are morphologically very similar. The only apparent difference is that the infectious particle of Ebola is somewhat longer. The infectious Ebola particle measures about 860 to 940 nm in length and the infectious Marburg particle, about 700 to 765 nm. In tissue culture preparations these viruses are highly variable in length. Some are several microns in length and branching is common, but they have a fairly constant diameter of about 70 nm.

REFERENCES

1. **Almeida, J. D., Howatson, A. F., Pinterci, L., and Fenje, P.,** Electron microscope observations on rabies virus by negative staining, *Virology,* 18, 147, 1962.
2. **Bergold, G. H. and Munz, K.,** Ultrastructure of Cocal, Indiana, and New Jersey serotypes of vesicular stomatitis virus, *J. Ultrastruct. Res.,* 17, 233, 1967.
3. **Bradish, C. J. and Kirkham, J. B.,** The morphology of vesicular stomatitis virus (Indiana C) derived from chick embryos or cultures of BHK 21/13 cells, *J. Gen. Microbiol.,* 44, 359, 1966.
4. **Emerson, S. U. and Wagner, R. R.,** Dissociation and reconstitution of the transcriptase and template activities of vesicular stomatitis B and T virions, *J. Virol.,* 10, 297, 1972.
5. **Howatson, A. F.,** Vesicular stomatitis and related viruses, *Adv. Virus Res.,* 16, 195, 1970.
6. **Howatson, A. F. and Whitmore, G. F.,** The development and structure of vesicular stomatitis virus, *Virology,* 16, 466, 1962.

Marburg-Ebola Viruses

7. **Ellis, D. S., Simpson, D. I. H., Francis, D. P., Knobloch, J., Bowen, E. T. W., Lolick, P., and Deng, I. M.,** Ultrastructure of Ebola virus particles in human liver, *J. Clin. Pathol.,* 31, 201, 1978.

8. **Johnson, K. M., Webb, P. A., Lange, J. V., and Murphy, F. A.,** Isolation and partial characterization of a new virus causing acute haemorrhagic fever in Zaire, *Lancet,* 1, 569, 1977.

9. **Malherbe, H. and Strickland-Cholmley, M.,** Human disease from monkeys (Marburg virus), *Lancet,* 1, 1434, 1968.

10. **Martini, G. A. and Siegert, R., Eds.,** *Marburg Virus Disease,* Springer-Verlag, Berlin, 1971.

11. **Murphy, F. A., Van der Groen, G., Whitfield, S. G., and Lange, J. V.,** Ebola and Marburg virus morphology and taxonomy, in *Ebola Virus Haemorrhagic Fever,* Pattyn, S. R., Ed., Elsevier/North-Holland Biomedical Press, New York, 1978, 61.

12. **Siegert, R.,** Marburg virus, *Virol. Monogr.,* 11, 97, 1972.

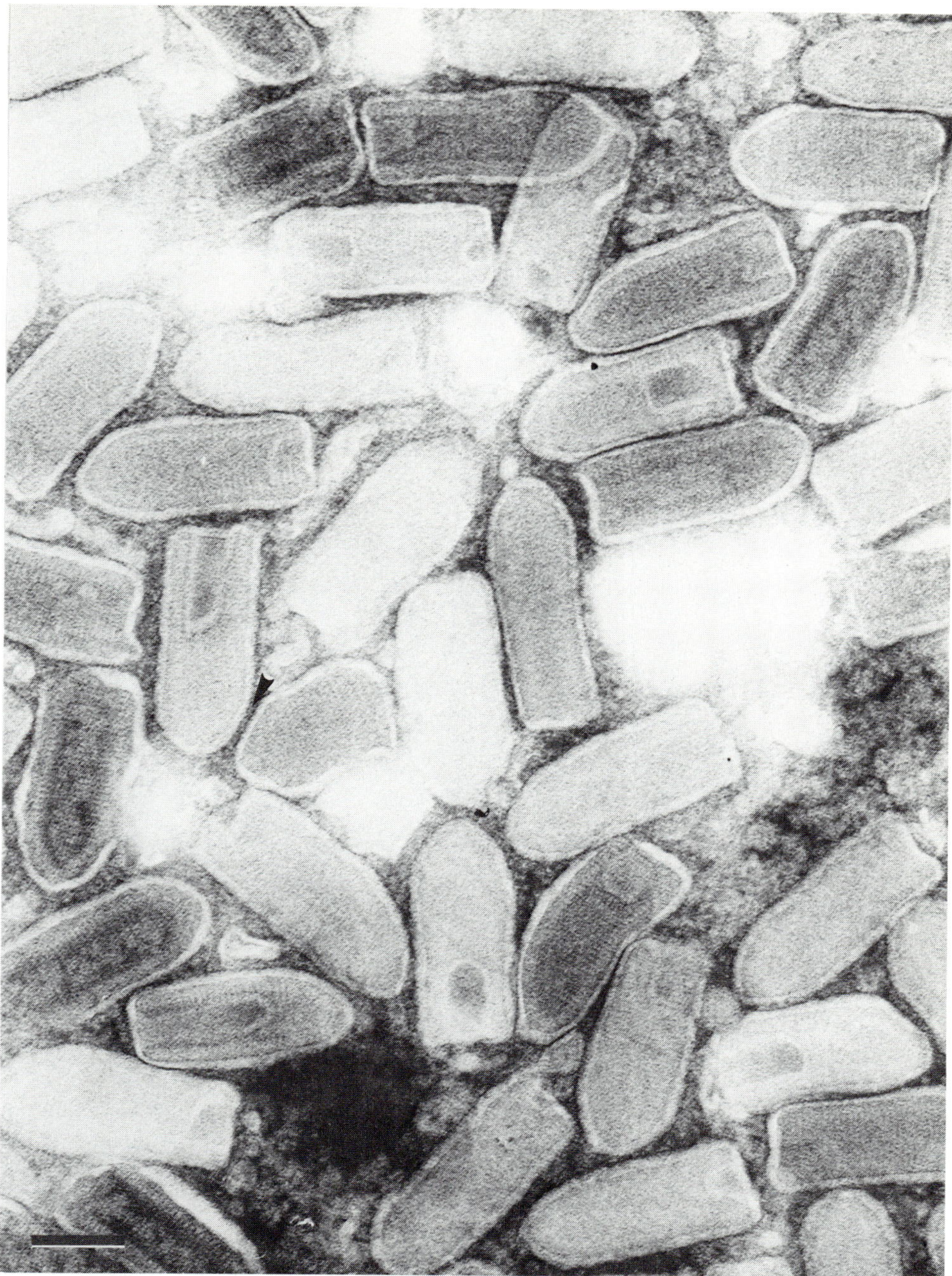

PLATE 1.   Purified preparation of vesicular stomatitis virus. The bullet shape is obvious as is the helically wound nucleocapsid. A layer of surface projections can also be seen on particles which are not closely packed together. All bars equal 100 nm.

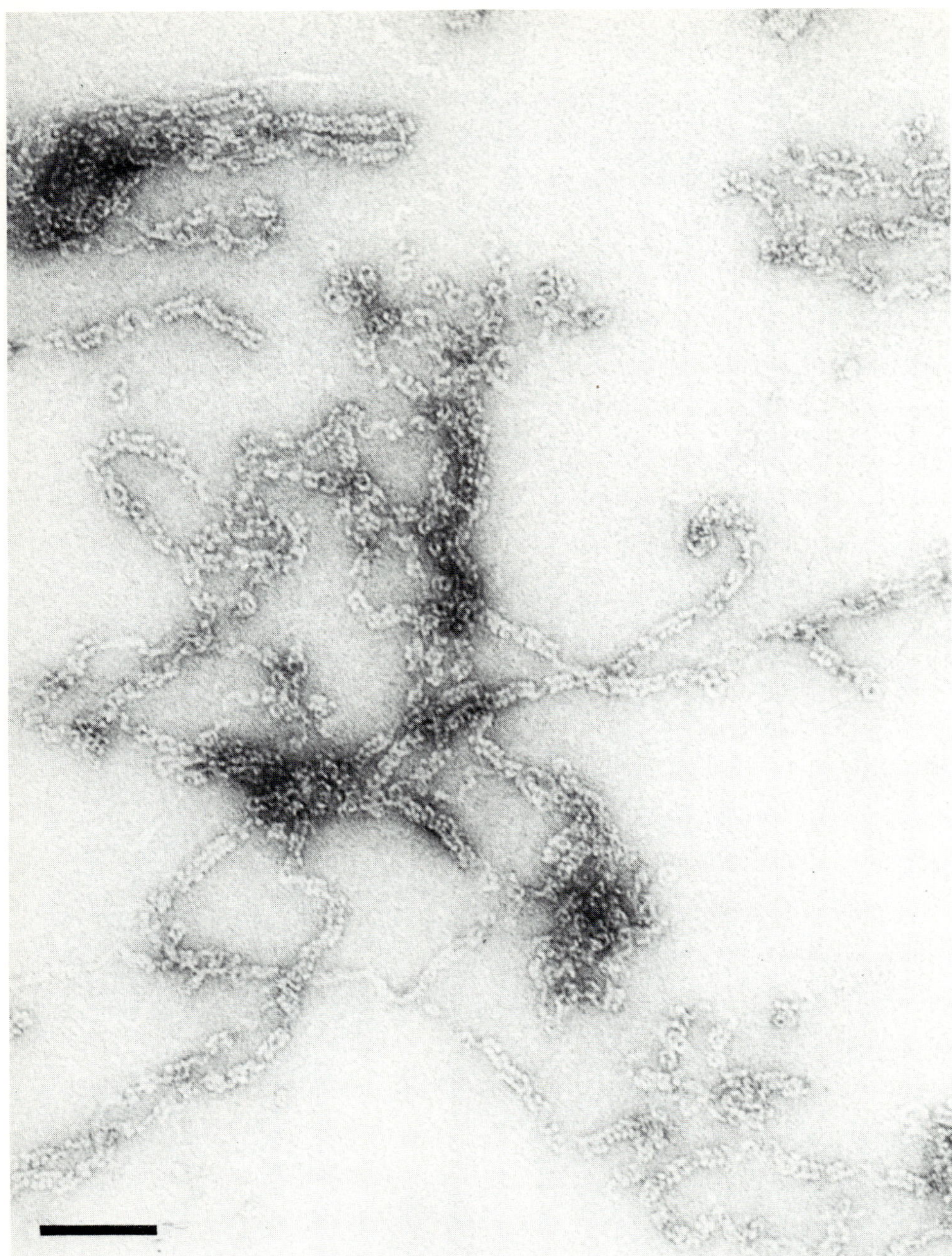

PLATE 2.   Ribbons of free-lying rabies virus ribonucleoprotein. It appears as a helix with protein radially arrayed around RNA. Strand length and number of turns of the helix vary among animal rhabdoviruses.

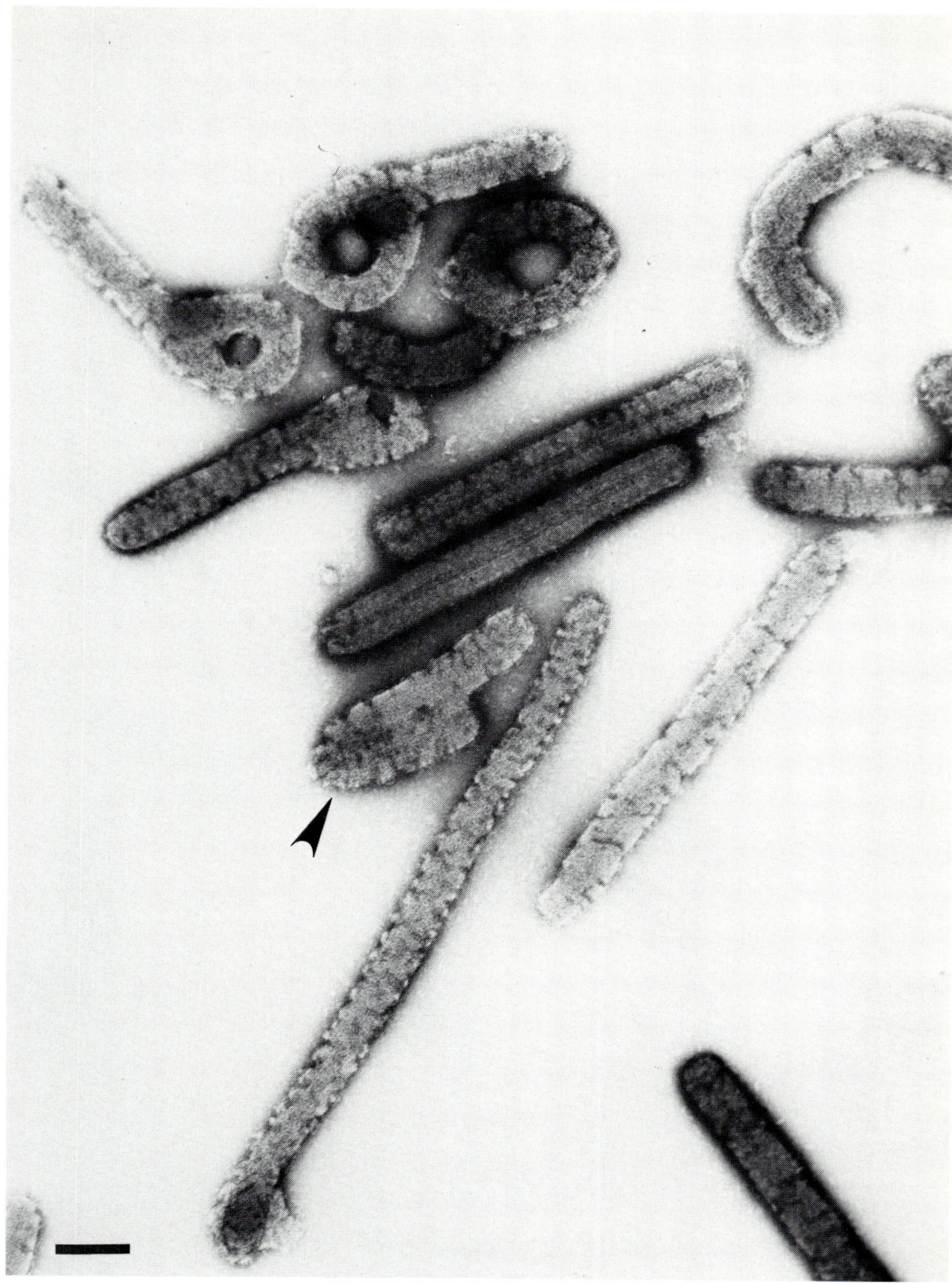

PLATE 3.   Marburg virus. Virus is frequently seen in form of 6's and 9's and in "hairpin" array (arrow). Cylindrical forms have a relatively wide axial channel and evenly spaced surface projections.

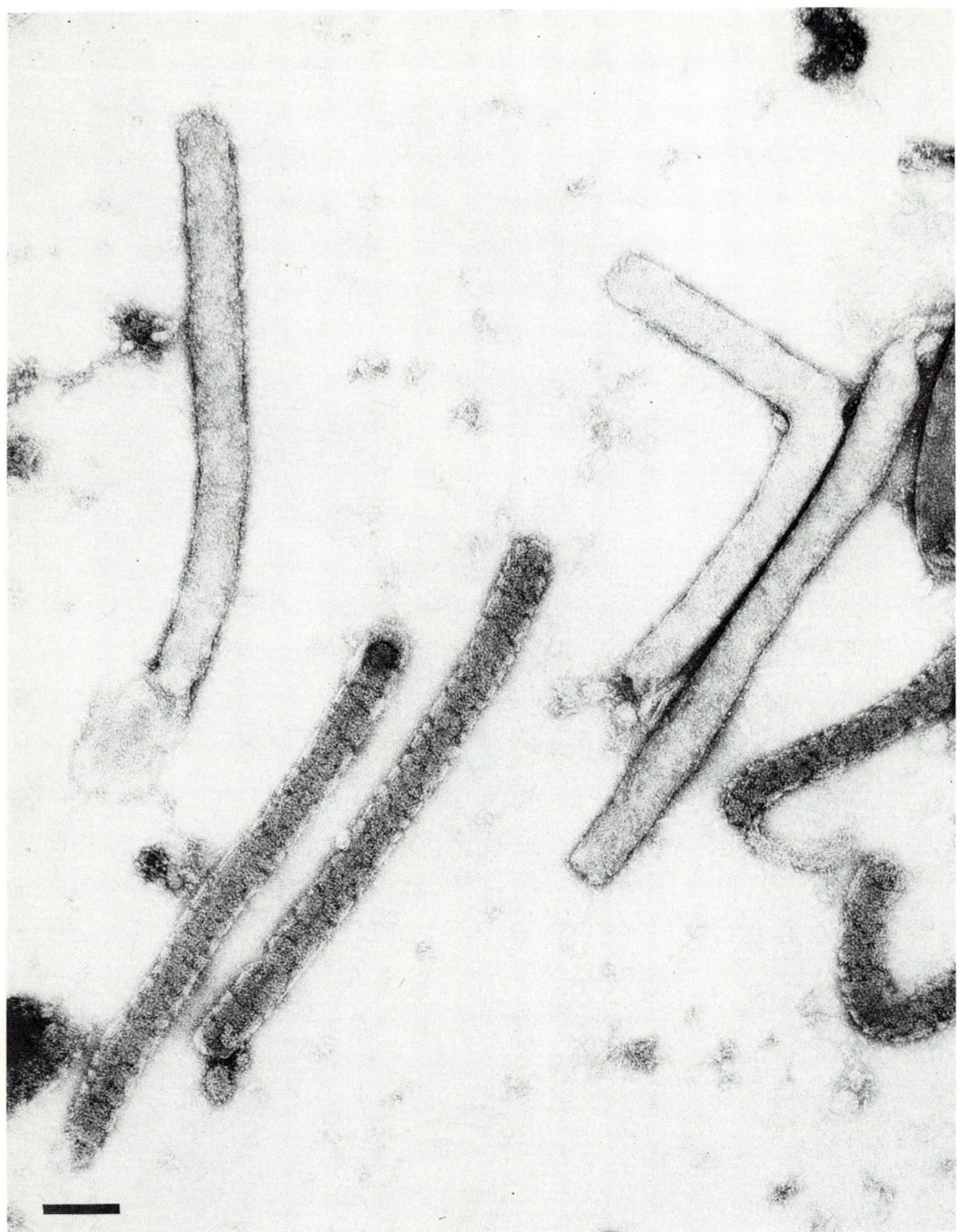

PLATE 4.   Infectious form of Ebola virus. Particles are relatively uniform in length and width. An envelope around particles and surface projections are evident, as is a wide axial channel.

Part IV
RNA Viruses with
Unknown Capsid Symmetry and
Complex Envelopes

Chapter 12

BUNYAVIRIDAE

The family Bunyaviridae is a taxonomic group of arboviruses which includes all former members of Bunyamwera supergroup and some previously unclassified arboviruses. About 90 viruses which had been assigned to 11 different serologically related arbovirus groups were brought together on the basis of serological cross-reactions to form the Bunyamwera supergroup named after the first recognized member, an arbovirus from the Bunyamwera area of Uganda. Subsequently, representative viruses of the supergroup were found to be morphologically indistinguishable from each other yet were distinct from all other established arbovirus groups. These viruses constitute the present genus *Bunyavirus.* An additional 54 viruses have now been found to be morphologically similar to members of the genus *Bunyavirus* but are serologically independent of the supergroup. The taxonomic status of these last viruses has not been clearly resolved at the time of this writing, that is, whether to include them in the genus *Bunyavirus* or to place them into one or more separate genera.

Well-studied members of Bunyaviridae such as LaCrosse and Uukuniemi virus have an average diameter of 92 to 105 nm. They are relatively homogeneous in appearance and have a unit membrane envelope 10- to 12-nm thick with radiating projections 10- to 12-nm long. The nucleocapsid is segmented and is of at least three size modes. The RNA is also tripartite and closed by bonding of complementary base pairs forming ''panhandles.''

Uukuniemi virus has projections which are clustered to form hollow, cylindrical morphological units which are 8- to 10-nm long and 10- to 12-nm in diameter with a 5 nm cavity. These projections are clustered into a penton-hexon T = 12 (P = 3) icosahedral lattice.

REFERENCES

1. **Bishop, D. H. L. and Shope, R. E.,** Bunyaviridae, in *Comprehensive Virology,* Vol. 14, Fraenkel-Conrat, H. and Wagner, R. R., Eds., Plenum Press, New York, 1979, 1.
2. **Holmes, I. H.,** Morphological similarity of Bunyamwera supergroup viruses, *Virology,* 43, 708, 1971.
3. **Murphy, F. A., Harrison, A. K., and Tzianabos, T.,** Electron microscopic observations of mouse brain infected with Bunyamwera group arboviruses, *J. Virol.,* 2, 1315, 1968.
4. **Murphy, F. A., Whitfield, S. G., Coleman, P. H., Calisher, C. H., Rabin, E. R., Jenson, A. B., Melnick, J. L., Edwards, M. R., and Whitney, E.,** California group arboviruses: electron microscopic studies, *Exp. Mol. Pathol.,* 9, 44, 1968.
5. **Murphy, F. A., Harrison, A. K., and Whitfield, S. G.,** Bunyaviridae: morphologic and morphogenetic similarities of Bunyamwera serologic super-group viruses and several other arthropod-borne viruses, *Intervirology,* 1, 297, 1973.
6. **Obijeski, J. F., Bishop, D. H. L., Palmer, E. L., and Murphy, F. A.,** Segmented genome and nucleocapsid of LaCrosse virus, *J. Virol.,* 20, 664, 1976.
7. **Obijeski, J. F. and Murphy, F. A.,** Bunyaviridae: recent biochemical developments, *J. Gen. Virol.,* 37, 1, 1977.
8. **Von Bonsdorff, C-H. and Pettersson, R.,** Surface structure of Uukuniemi virus, *J. Virol.,* 16, 1296, 1975.

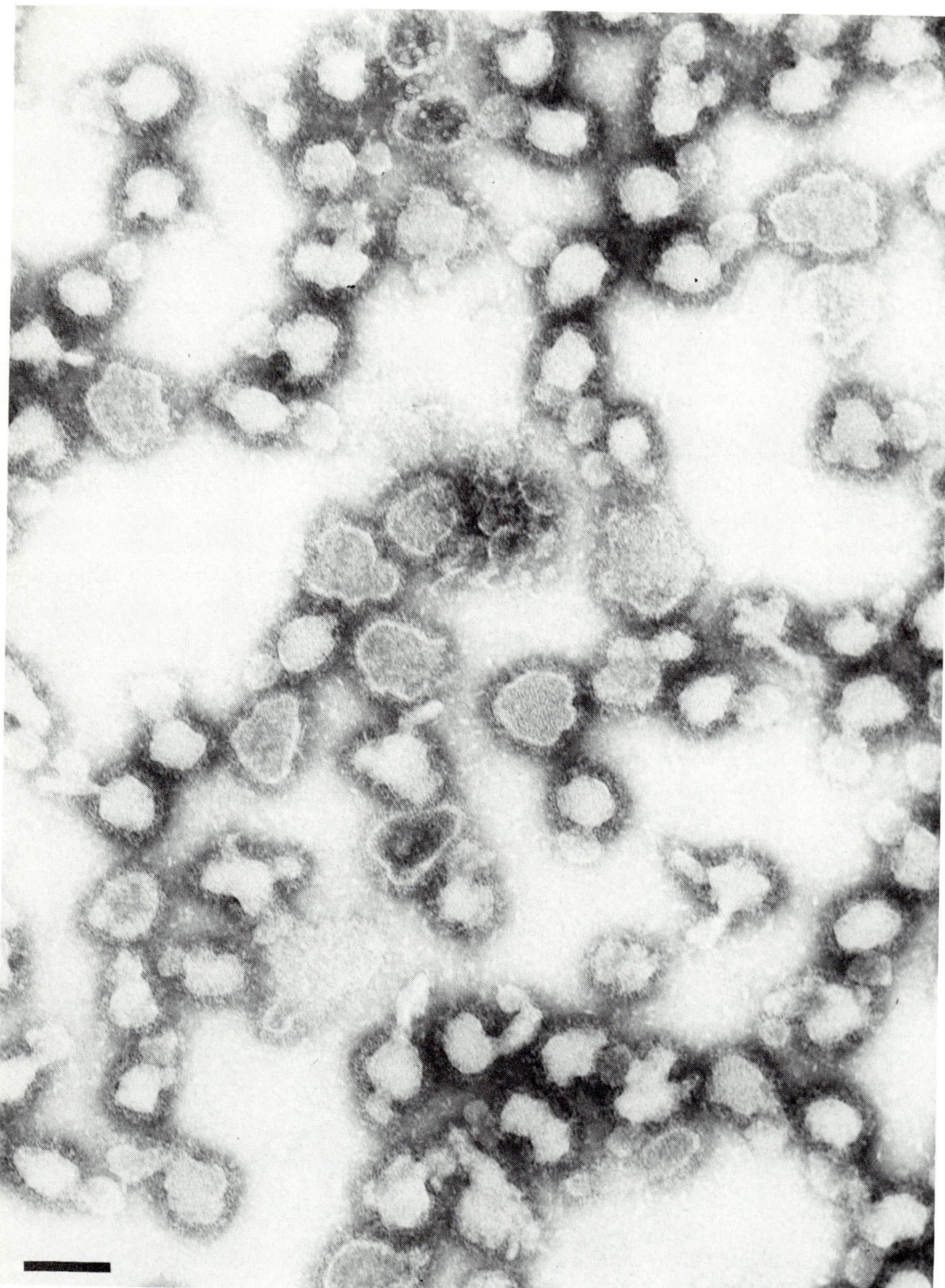

PLATE 1. LaCrosse virus. Surface projections 10- to 12-nm long radiate from a unit membrane envelope. The projections do not appear to be clustered into a symmetrical array. Particles are somewhat
pleomorphic and variable in diammeter. All bars equal 100 nm.

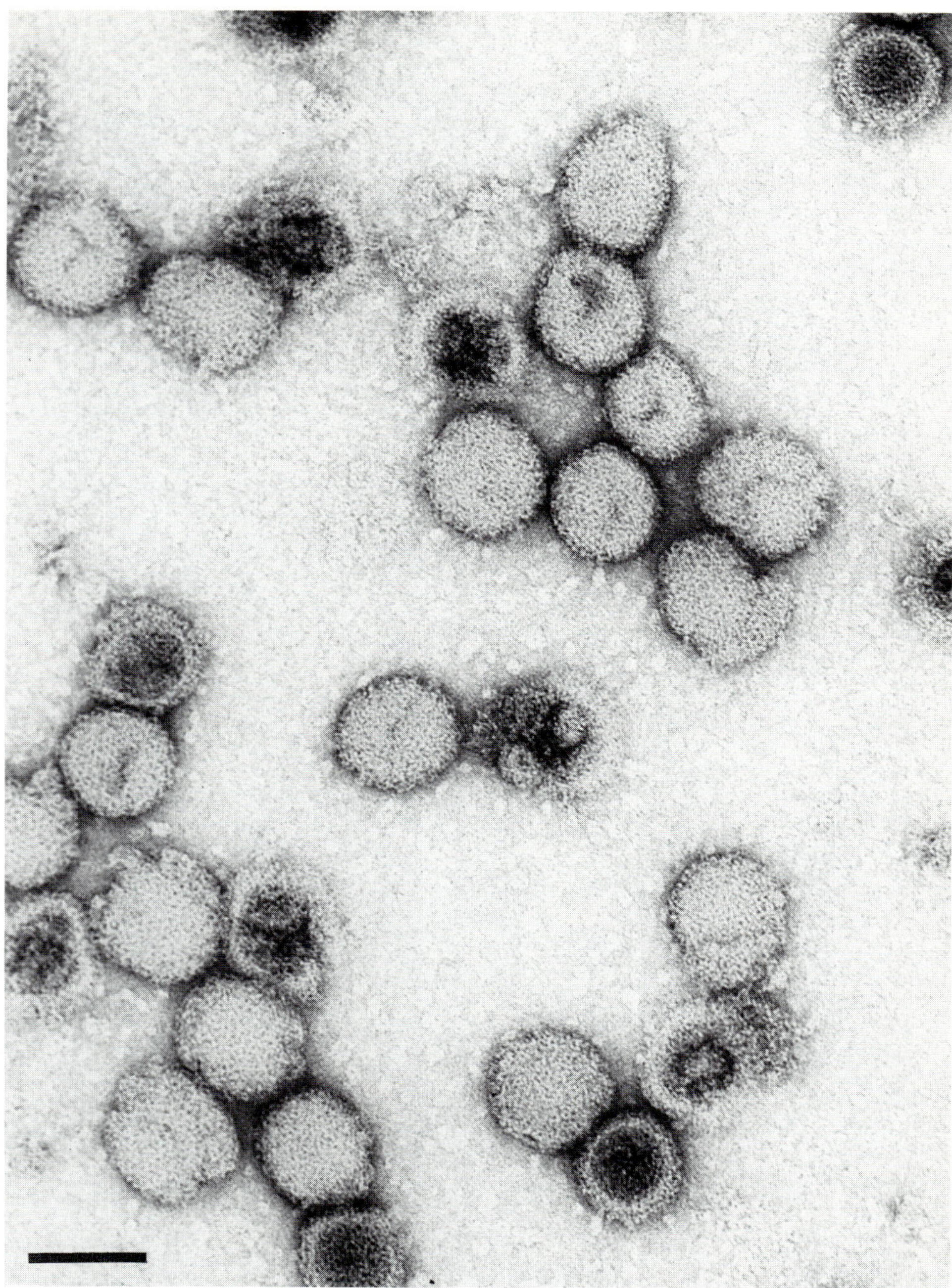

PLATE 2. Uukuniemi virus. This virus has projections which are clustered to form hollow cylindrical units. They are clustered into a penton-hexon T = 12 (P = 3) iscosahedral lattice.

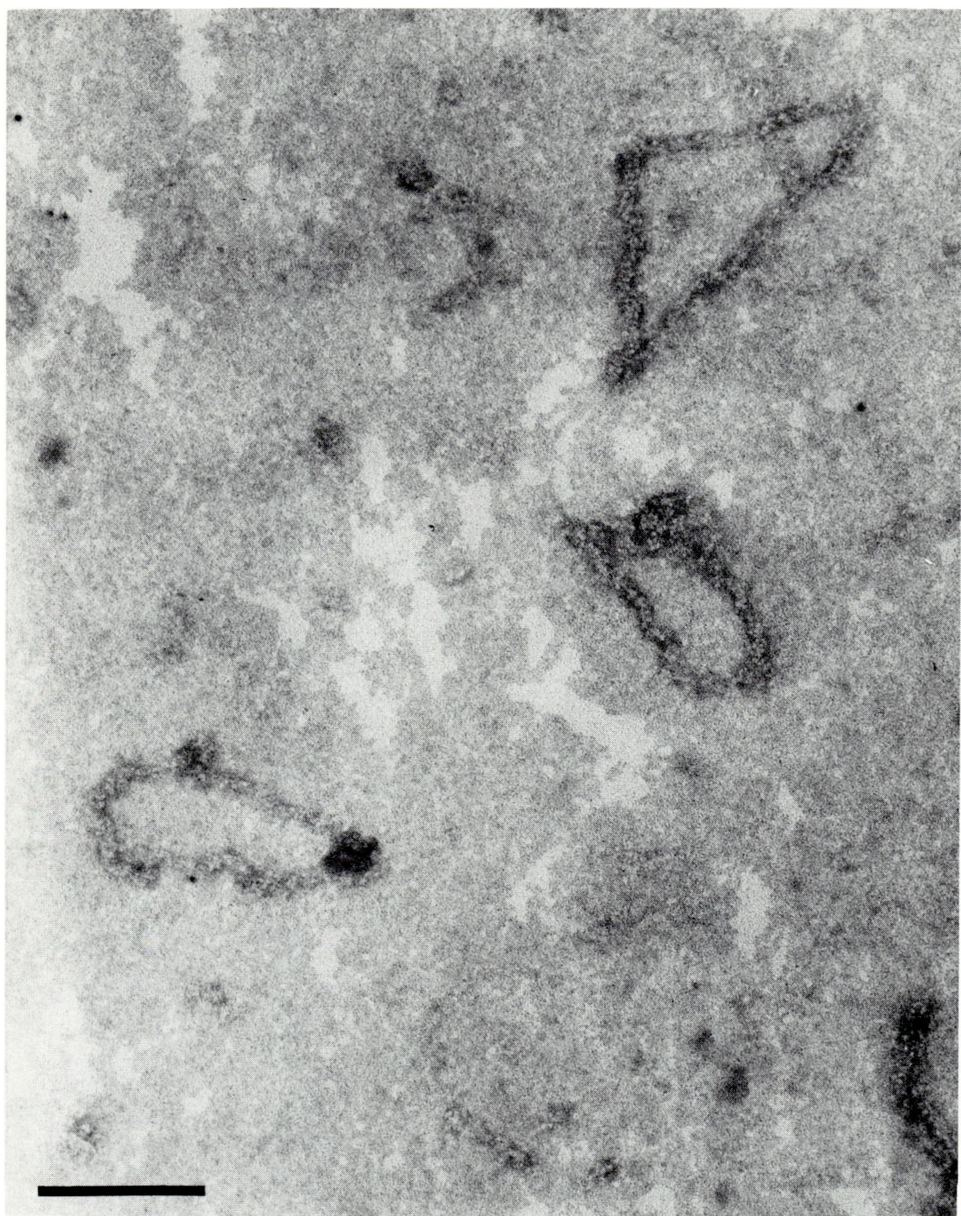

PLATE 3. Circular nucleocapsids of LaCrosse virus. They are convoluted and supercoiled. There are three size modes: 200, 510, and 700 nm. Shown here is the intermediate 510 nm mode.

Chapter 13

ARENAVIRIDAE

There are currently 10 known viruses of the family Arenaviridae which fall into 3 groups: lymphocytic choriomeningitis virus (LCM); the Tacaribe complex which includes Junin (Argentinian hemorrhagic fever), Machupo (Bolivian hemorrhagic fever), Amapari, Tacaribe, Pichinde, Parana, Latino, and Tamiami viruses; and Lassa virus. Arenaviruses share a common group antigen but do not cross-neutralize. Lassa virus and some of the other arenaviruses which cause hemorrhagic fever are extremely hazardous and are currently studied in only a few specialized laboratories with special safety precautions. The only known arenaviruses found in the U.S. are LCM and Tamiami viruses. The latter is not known to be pathogenic for man.

The name arenavirus is derived from the characteristic host-cell ribosomes, which appear as fine granules (arena = sand) seen within virions by thin-section electron microscopy. The structures are diagnostic for arenaviruses. Diagnosis by negative-stain EM has not been reported, but the virus does have distinctive morphology and can readily be distinguished from other known viruses by this method.

All arenaviruses have the same morphology. Virions range from 50 to 300 nm in diameter and are spherical or pleomorphic. They have an envelope which is sensitive to lipid solvents and from which club-like projections 10 nm in length radiate. These projections are closely spaced and have not been seen in any clearly defined pattern. The core of arenaviruses contains several pieces of single-stranded RNA, two of which are virus-specific. Together with protein, this RNA forms circular ribonucleoprotein of at least two size modes of 650 and 1300 nm in length. The core also contains host-cell ribosomal RNA and protein. The Arenaviridae and some members of the Bunyaviridae are the only viruses known to have circular, segmented ribonucleoprotein as nucleocapsids.

REFERENCES

1. World Health Organization, International symposium on arenavirus infections of public health importance, *Bull. WHO,* 52, 4, 1975.
2. **Dalton, A. J., Rowe, W. P., Smith, G. H., Wilsnack, R. E., and Pugh, W. E.,** Morphological and cytochemical studies on lymphocytic choriomeningitis virus, *J. Virol.,* 2, 1465, 1968.
3. **Mannweiler, K. and Lehmann-Grube, F.,** Electron microscopy of LCM virus-infected L cells, in *Lymphocytic Choriomeningitis Virus and Other Arenaviruses,* Lehmann-Grube, F., Ed., Springer Verlag, Berlin, 1973, 37.
4. **Murphy, F. A., Webb, P. A., Johnson, K. M., and Whitfield, S. G.,** Morphological comparison of Machupo with lymphocytic choriomeningitis virus: basis for a new taxonomic group, *J. Virol.,* 4, 535, 1969.
5. **Murphy, F. A. and Whitfield, S. G.,** Morphology and morphogenesis of arenaviruses, *Bull. WHO,* 52, 409, 1975.
6. **Palmer, E. L., Obijeski, J. F., Webb, P., and Johnson, K. M.,** The circular, segmented nucleocapsid of an arenavirus—Tacaribe virus, *J. Gen. Virol.,* 36, 541, 1977.
7. **Rawls, W. E. and Leung, W. C.,** Arenaviruses, in *Comprehensive Virology,* Vol. 14, Fraenkel-Conrat, H. and Wagner, R. R., Eds., Plenum Press, New York, 1979, 157.
8. **Spier, R. W., Wood, O., Liebheber, H., and Buckley, S. M.,** Lassa fever, a new virus disease of man from West Africa. IV. Electron microscopy of Vero cell cultures infected with Lassa virus, *Am. J. Trop. Med. Hyg.,* 19, 692, 1970.

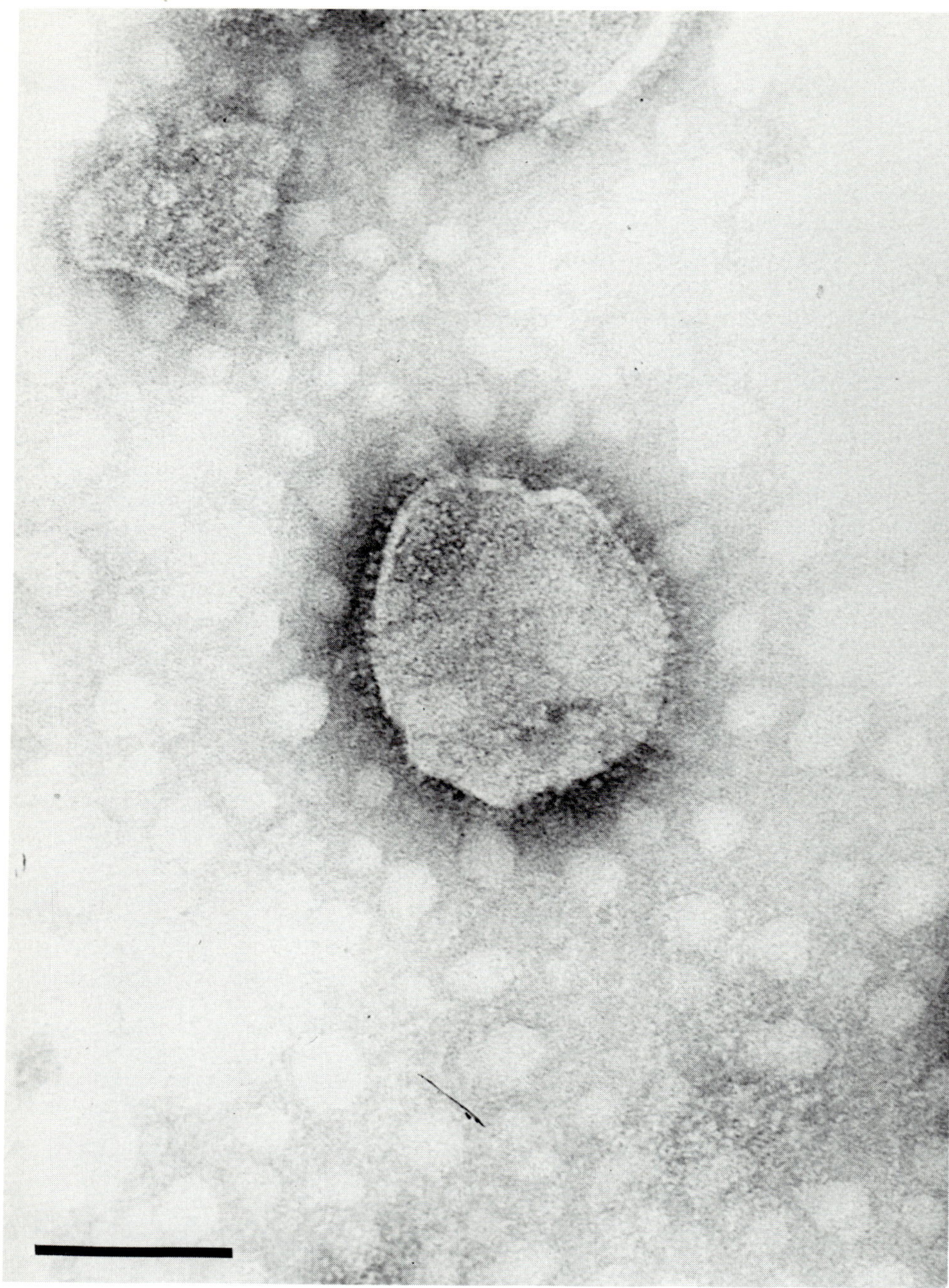

PLATE 1. Tacaribe virus. The particle has a well defined membrane from which large club-shaped surface projections radiate. The virion is roughly circular. All arenaviruses have the same morphology. All bars equal 100 nm.

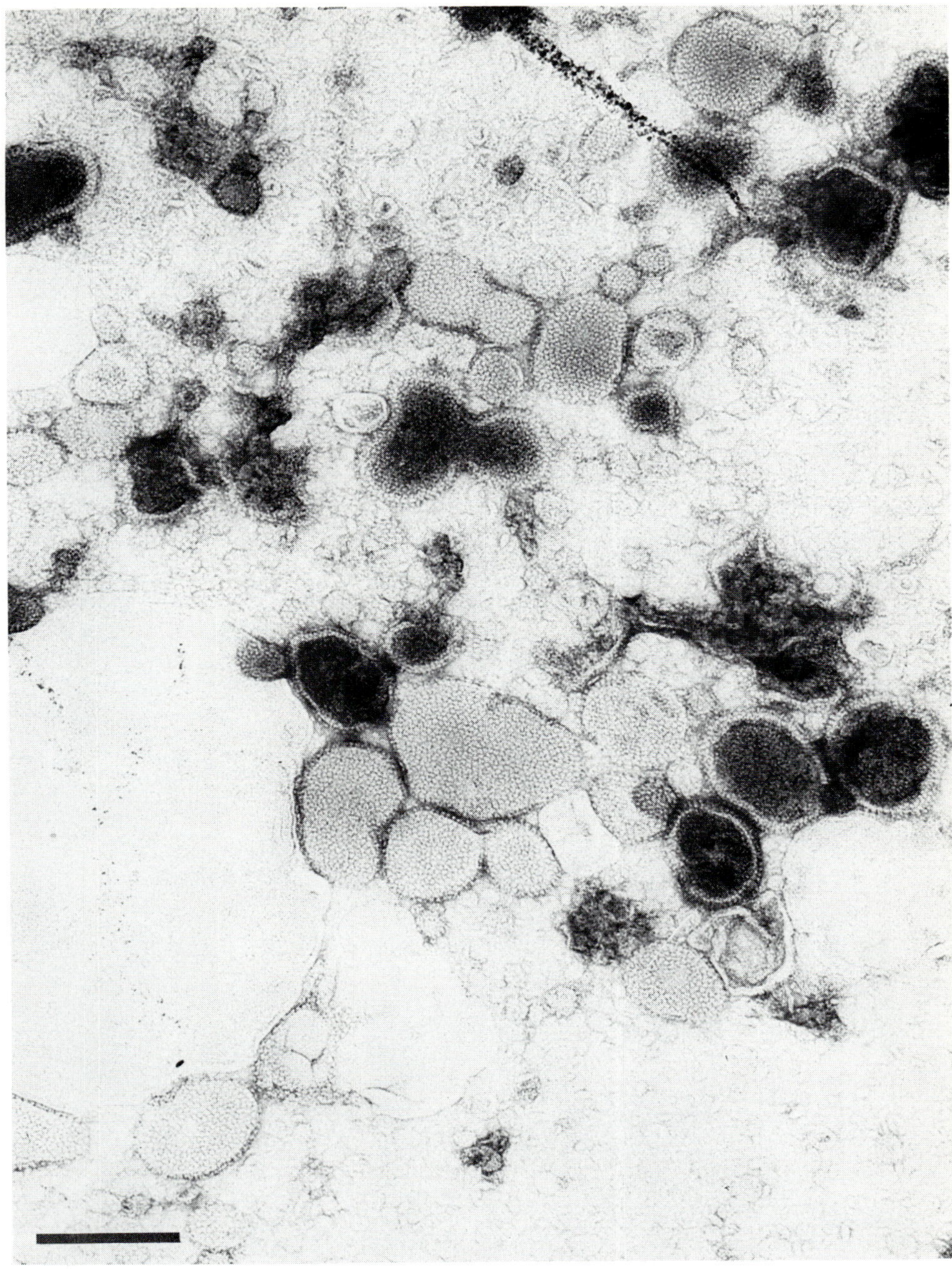

PLATE 2. Tacaribe virus. Particles are viewed head-on, so that surface projections appear hexagonal with an axial hole. Arenaviruses are variable (50 to 300 nm) in diameter.

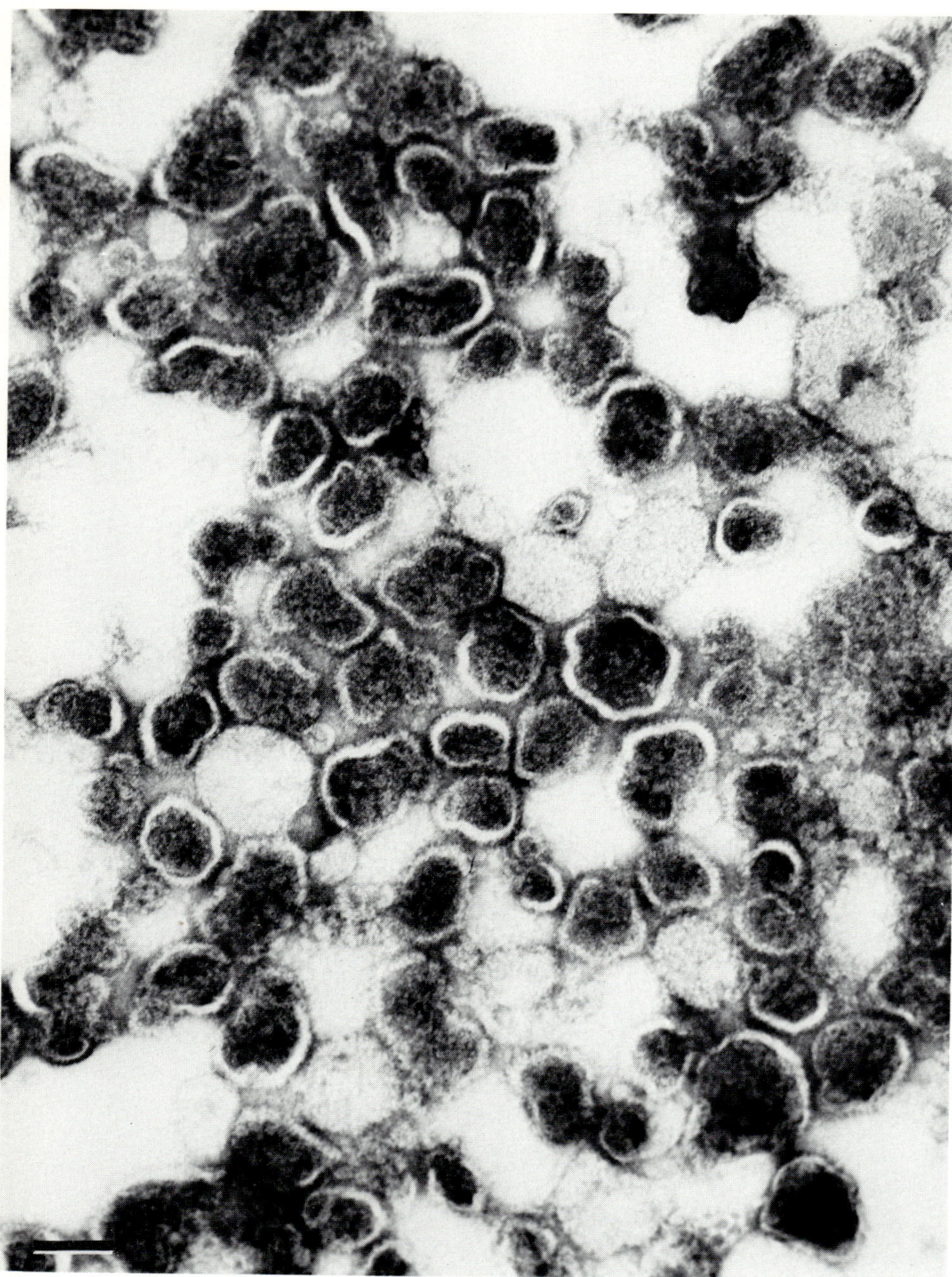

PLATE 3. Tacaribe virus. Appearance of Tacaribe virus after extensive purification in CsCl density gradients. UA stain has penetrated the viral envelope to reveal coils of electron-dense ribonucleoprotein surrounded by an electron-lucent envelope.

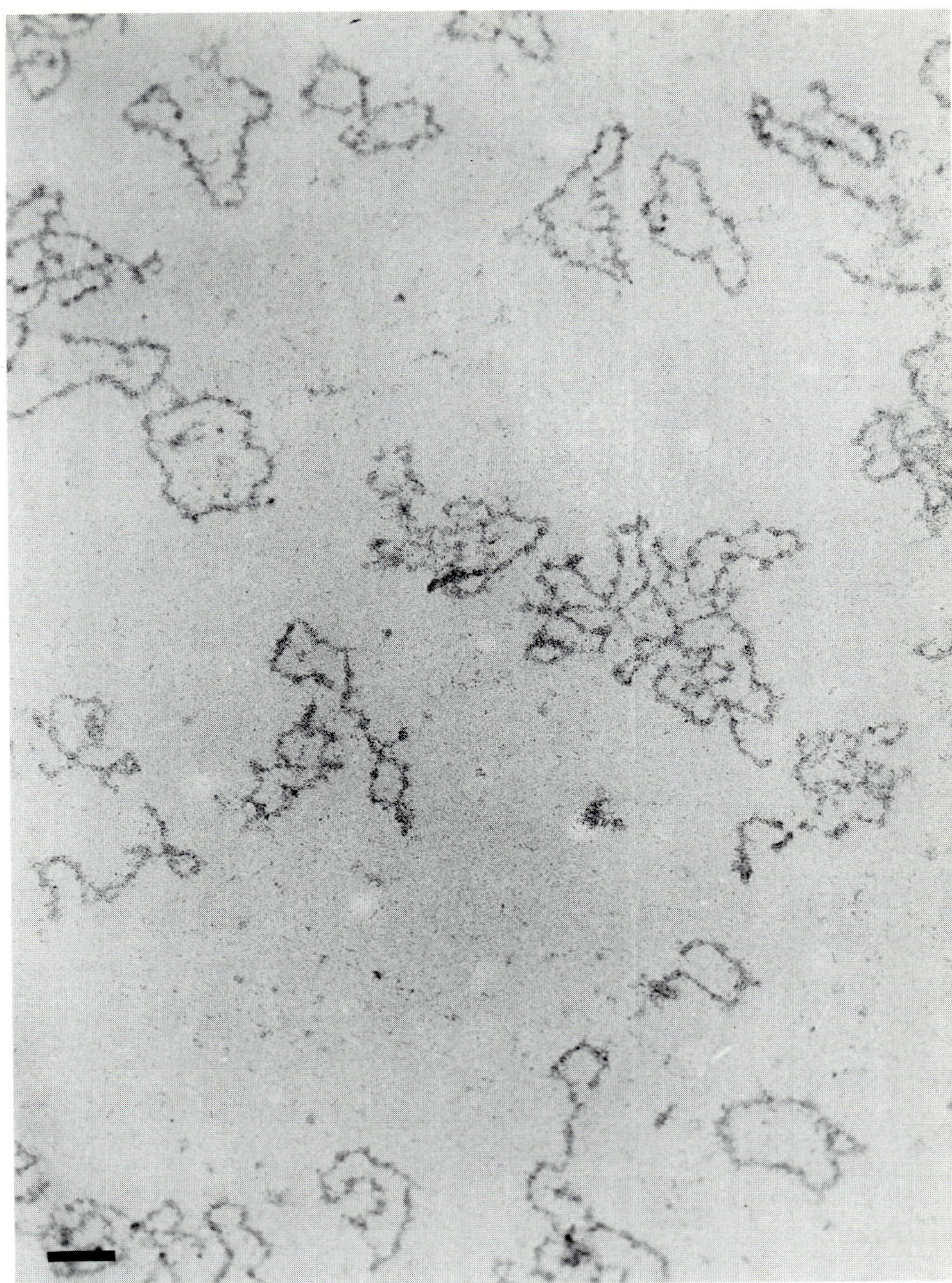

PLATE 4. Survey micrograph of Tacaribe RNP from virions lysed with nonionic detergent. Most forms are joined end to end.

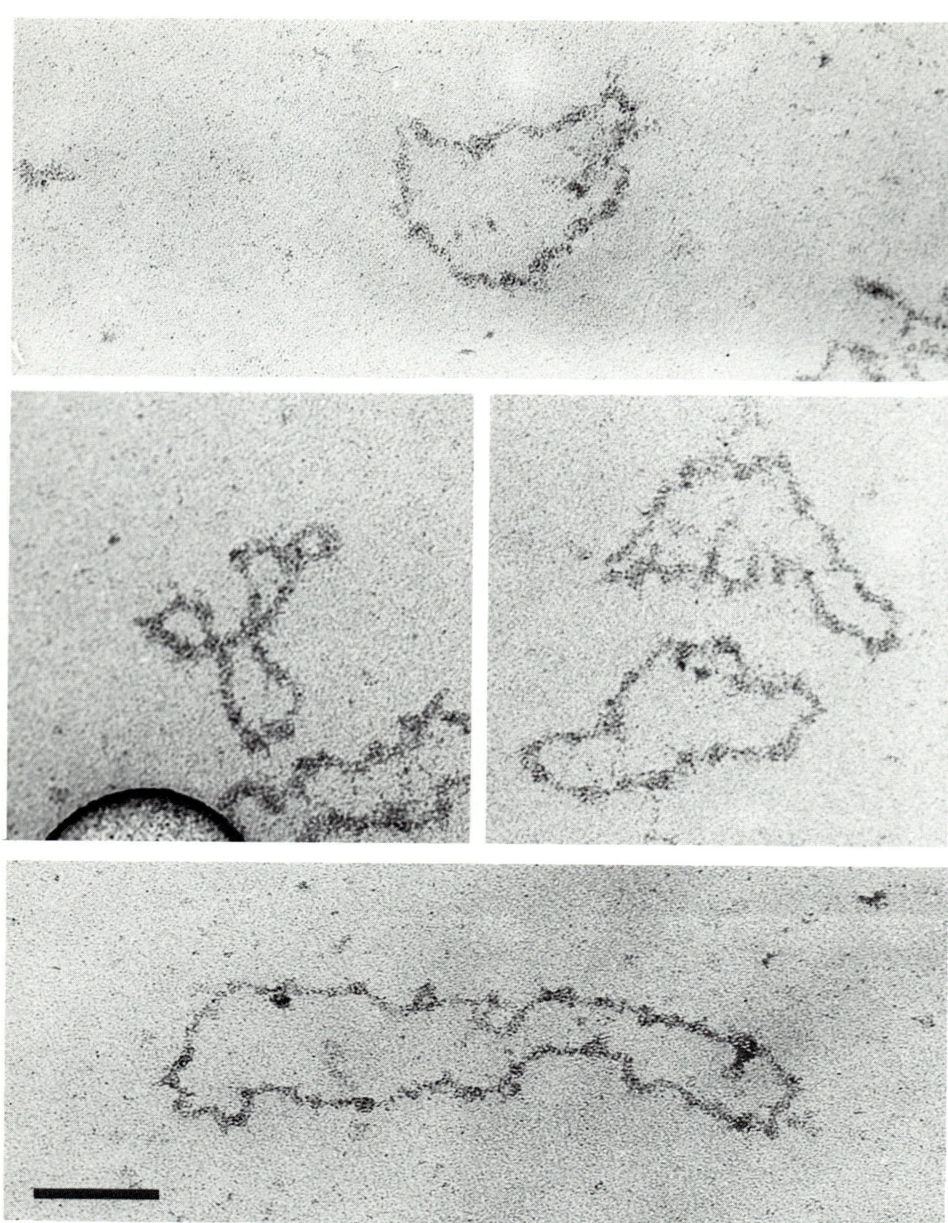

PLATE 5. Two predominant forms of Tacaribe virus circular ribonucleoprotein. One is about twice the size of the other. They are roughly 1300 nm and 650 nm in length.

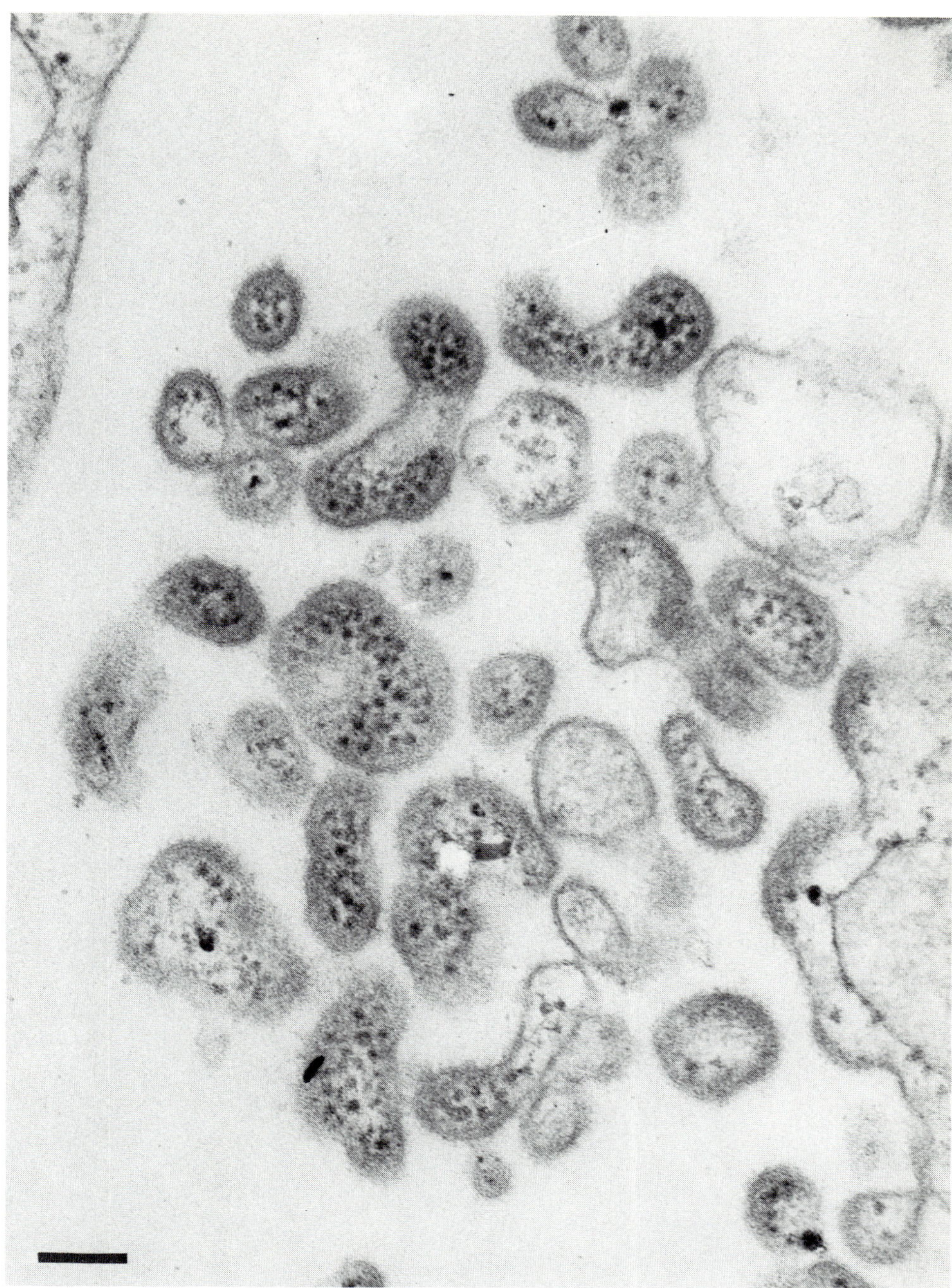

PLATE 6. Thin-section preparation of LCM virus in VERO cells. Electron-dense ribosomes inside virions are diagnostic for the Arenaviridae. These structures are not seen by negative-stain EM.

Chapter 14

CORONAVIRIDAE

Coronaviruses (family Coronaviridae) of importance to humans are recovered primarily from adults with upper respiratory tract illness. However, some reports indicate that these viruses can be involved in exacerbation of chronic lower respiratory tract disease as well. Coronaviruses have also been recovered or visualized from a host of lower animals, including murine, porcine, avian, bovine, canine, feline, and equine species. They cause a wide spectrum of illnesses in such animals, ranging from hepatitis in mice to diarrhea in calves. Strains that infect lower animals are not thought to cause disease in humans.

The human coronaviruses include at least three distinct but antigenically related prototype strains: B814, 229E, and OC43. Several uncharacterized subtypes have also been isolated from the human respiratory tract. Coronaviruses are fastidious in their growth requirements in the laboratory and may be divided into two groups, namely, those which can be isolated only in organ culture and those which can be isolated in cell culture. B814 and OC43 coronaviruses were originally isolated in human embryonic tracheal organ cultures, and 229E was orginially isolated in human embryonic kidney cell culture. Subsequently, OC43 has been adapted to growth in suckling mouse brain and 229E in WI-38 fibroblasts and other cell cultures.

Coronavirus OC43 possesses a hemagglutinin, but unlike *Influenza* A and B, it does not have a neuraminidase. Although a hemagglutinin has not yet been demonstrated in other human strains thus far isolated, certain other animal strains possess a hemagglutinin. Coronaviridae infectivity is labile to pH, heat, and treatment with lipid solvents.

Coronaviruses are generally spherical, with some pleomorphism. They range from about 80 to 130 nm or more in diameter. Electron micrographs of thin sections show an outer envelope about 8 nm thick surrounding an electron-dense shell and a central zone enclosing a genome that is probably arranged as a loosely wound single-stranded RNA protein helix. Examination of coronavirus preparations by negative-stain EM, however, usually reveals moderately pleomorphic particles covered with projections, known as peplomers. These projections are quite distinctive and give virions the appearance of a solar corona (hence the name coronavirus). The projections have a narrow base, and are 12- to 24 nm long. Their ends are club-shaped with an approximate diameter of about 10 nm. Symmetry to their arrangement has not as yet been noted, although when one views them end-on, hexagonally arrayed projections are sometimes seen. The envelope can sometimes be seen when stain penetrates partially disrupted particles.

REFERENCES

1. **Almeida, J. D., Berry, D. M., Cunningham, C. H., Hamre, D., Hofstad, M. S., Mallucci, L., McIntosh, K., and Tyrrell, D. A. J.**, Coronaviruses, *Nature (London)*, 220, 650, 1968.
2. **Becker, W. B., McIntosh, K., Dees, J. H., and Chanock, R. M.**, Morphogenesis of avian infectious bronchitis and related human virus (strain 229E), *J. Virol.*, 1, 1019, 1967.
3. **Berry, D. M., Cruikshank, J. G., Chu, H. P., and Wells, R. J. H.**, The structure of infectious bronchitis virus, *Virology*, 23, 403, 1964.
4. **Bingham, R. W. and Almeida, J. D.**, Studies on the structure of a coronavirus—avian infectious bronchitis virus, *J. Gen. Virol.*, 36, 495, 1977.
5. **Hamre, D., Kindig, D. A., and Mann, J.**, Growth and intracellular development of a new respiratory virus, *J. Gen. Virol.*, 1, 180, 1967.

6. **Hierholzer, J. C., Palmer, E. L., Whitfield, S. G., Kaye, H. S., and Dowdle, W. R.,** Protein composition of coronavirus OC43, *Virology,* 48, 516, 1972.

7. **Kapikian, A. Z.,** Coronaviruses, in *Diagnostic Procedures for Viral and Rickettsial Infections,* 4th ed., Lennette, E. H. and Schmidt, N. J., Eds., American Public Health Association, New York, 1969, 931.

8. **Kennedy, D. A. and Johnson-Lussenburg, C. M.,** Isolation and morphology of the internal component of human coronavirus, strain 229E, *Intervirology,* 6, 197, 1975/1976.

9. **McIntosh, K.,** Coronaviruses: a comparative review, *Curr. Top. in Microbiol. Immunol.,* 63, 85, 1974.

10. **Robb, J. A. and Bond, C. W.,** Coronaviridae, in *Comprehensive Virology,* Vol. 14, Fraenkel-Conrat, H. and Wagner, R. R., Eds., Plenum Press, New York, 1979, 193.

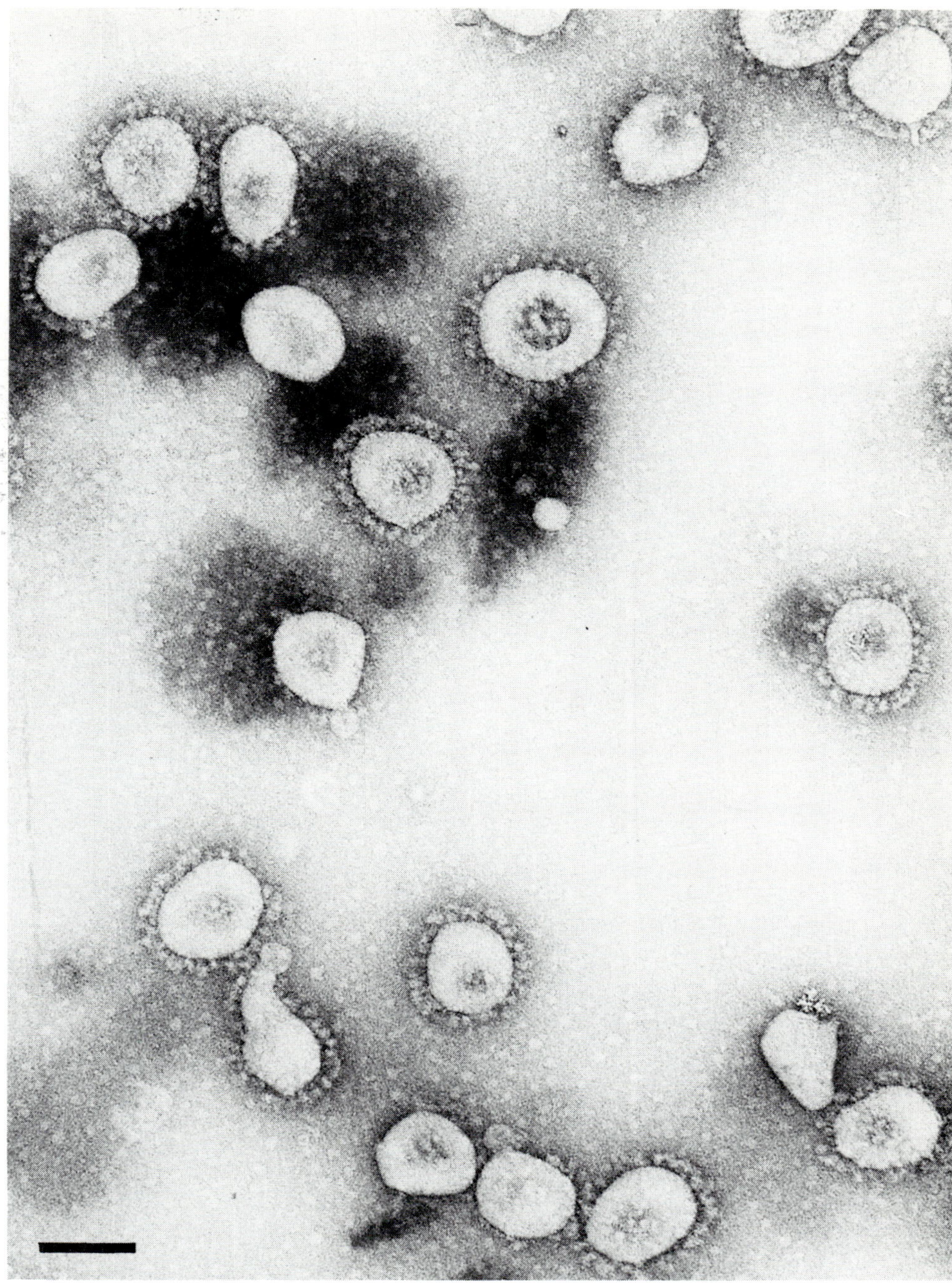

PLATE 1.   Coronavirus OC43. Virions are pleomorphic averaging 100 nm in diameter. They have unique club-shaped peplomers which project from the envelope and give the virus the appearance of a solar corona. Virus was purified from infected mouse brain. All bars equal 100 nm.

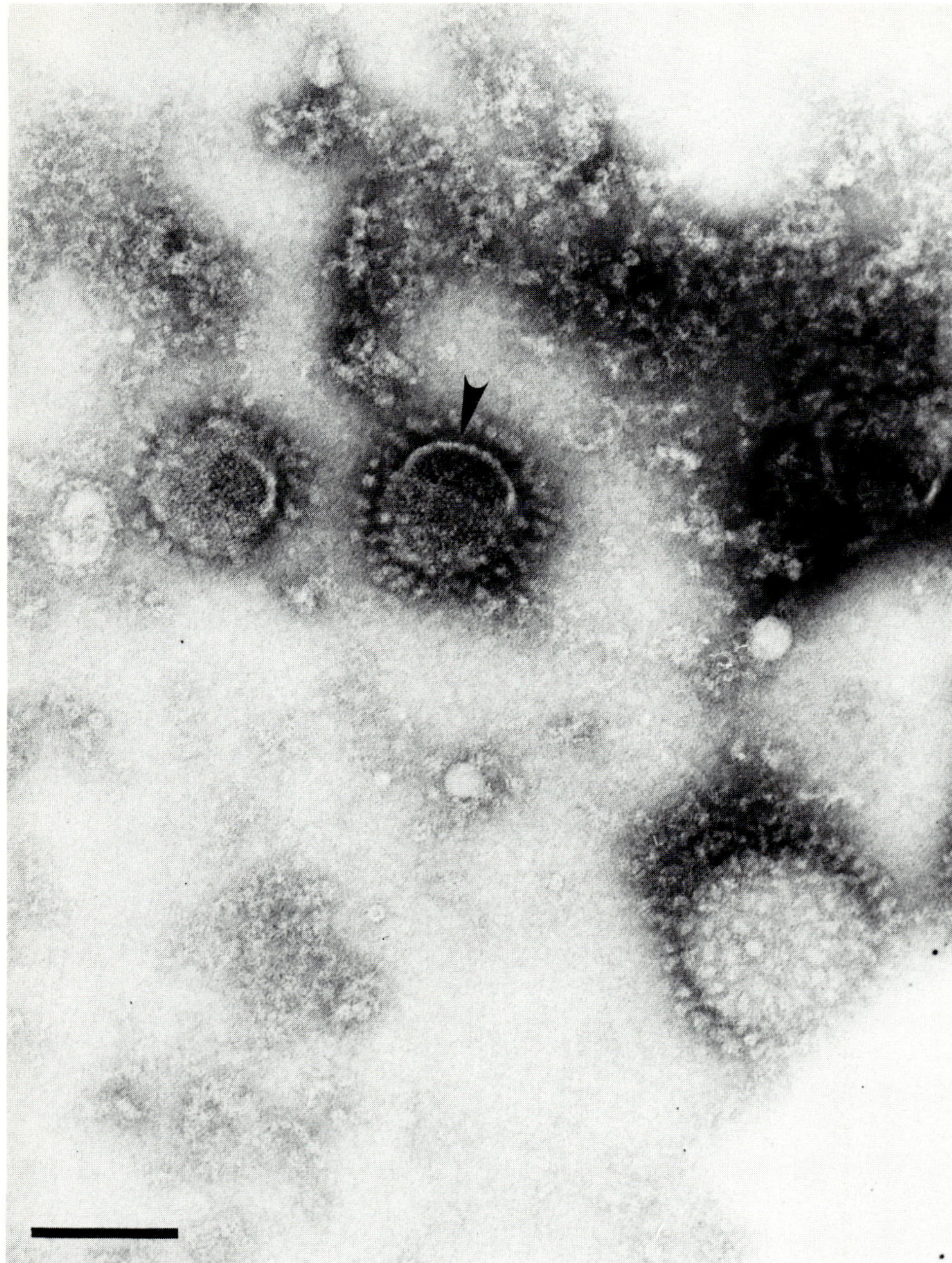

PLATE 2.   Partially disrupted coronavirus OC43 showing the relatively thick envelope (arrow) of the virus.

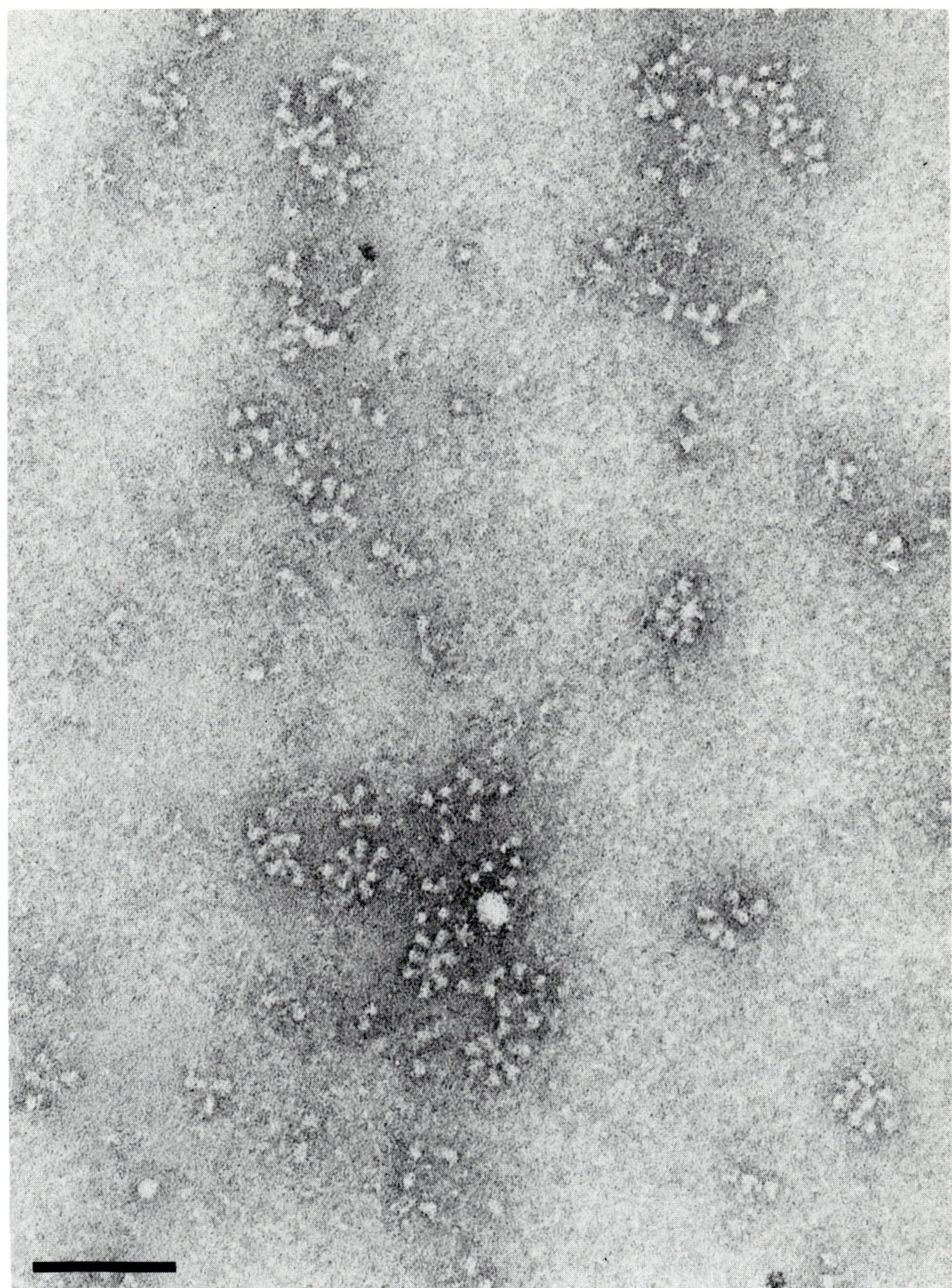

PLATE 3. Isolated surface projections of calf diarrhea coronavirus. The peplomers frequently array in rosettes and seem to have a small axial hole.

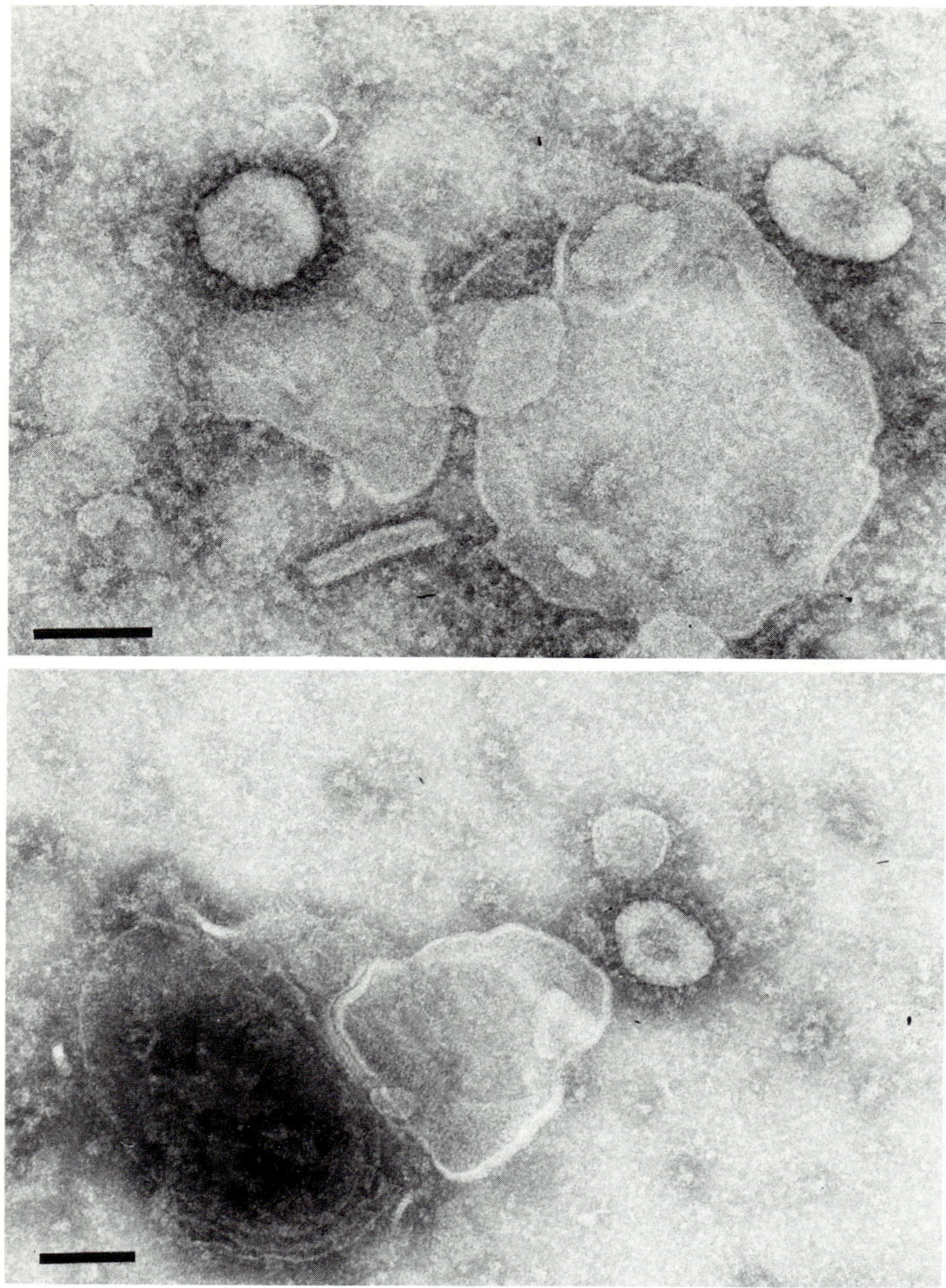

PLATE 4.   Coronavirus particles among tissue culture cell debris. Particles are readily recognized as *Coronaviridae* by the characteristic "solar corona" arrangement of surface projections.

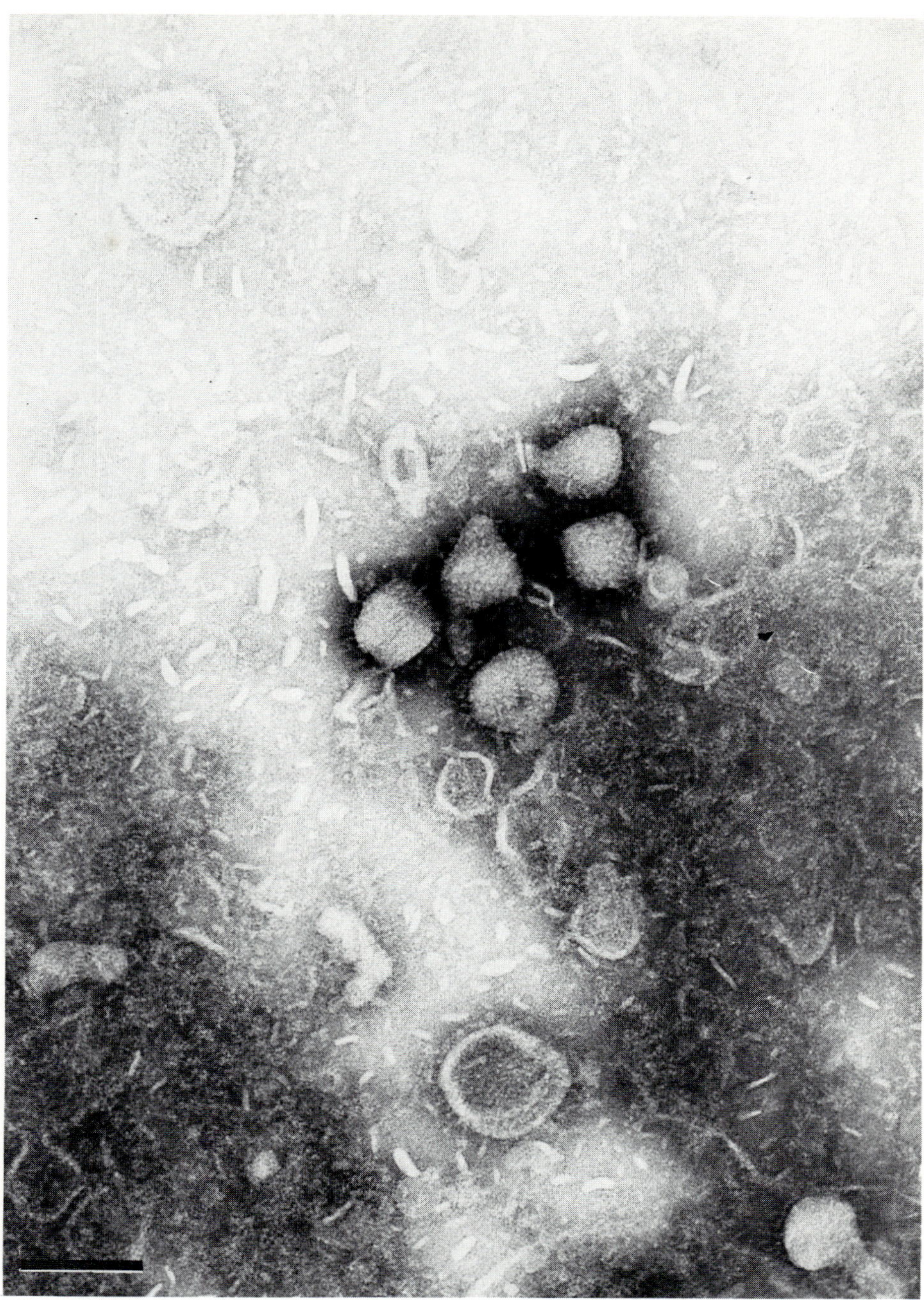

PLATE 5.   Coronavirus particles which are slightly distorted among mouse brain debris.

*Part V
DNA Viruses with
Cubic Capsid Symmetry
and Naked Nucleocapsids*

Chapter 15

PARVOVIRIDAE

The parvoviruses (parvus = small) are the smallest known viruses which infect vertebrates. Most studies suggest that they have an average diameter of about 22 nm. Virions have single-stranded DNA genomes. They are divided into two groups: (1) the adenovirus-associated viruses (AAV), which are nonproductive and require adenovirus for replication, and (2) a group of productive viruses such as hamster osteolytic virus (H-1) and feline panleukopenia virus, which can replicate autonomously.

The parvoviruses are generally not closely related antigenically, and the productive ones replicate best in rapidly dividing cells. Whether AAV causes disease has been obscured by their requirement for the presence of adenoviruses for replication. None of the known productive parvoviruses produce symptomatic disease in man, although particles described as "parvovirus-like," such as the Norwalk agent, have been implicated in gastroenteritis of humans. Feline panleukopenia virus, however, is an extremely important feline pathogen and may also be a pathogen for other animals. What appears to be a mutant strain of this virus has recently emerged as a highly lethal pathogen of dogs. Hepatitis A virus that affects humans is a possible member of the group.

The symmetry of parvovirus is definitely icosahedral. Current evidence, however, suggests that the capsid might be composed of either 12, 20, or even 32 capsomeres. The small size of these viruses has hindered definitive detailed structural analysis of the capsid by currently available techniques.

REFERENCES

1. **Bachmann, P. A., Hoggan, M. D., Melnick, J. L., Pereira, H. G., and Vago, C.,** Parvoviridae, *Intervirology,* 5, 83, 1975.
2. **Chandra, S. and Toolan, H. W.,** Electron microscopy of the H-1 virus. I. Morphology of the virus and a possible virus-host relationship, *J. Natl. Cancer Inst.,* 27, 1405, 1961.
3. **Dalton, A. J., Kilham, L., and Ziegler, R. E.,** A comparison of polyoma, "K", and Kilham rat viruses with the electron microscope, *Virology,* 20, 391, 1963.
4. **Hoggan, M. D.,** Adenovirus associated viruses, *Prog. Med. Virol.,* 12, 211, 1970.
5. **Hoggan, M. D.,** Small DNA viruses, in *Comparative Virology,* Maramorosch, K. and Kurstak, E., Eds., Academic Press, New York, 1971, 43.
6. **Karasaki, S.,** Size and ultrastructure of the H-viruses as determined with the use of specific antibodies, *J. Ultrastruct. Res.,* 16, 109, 1966.
7. **Mayor, H. D., Jamison, R. M., Jordan, L. E., and Melnick, J. L.,** Structure and composition of a small particle prepared from a simian adenovirus, *J. Bacteriol.,* 90, 235, 1965.
8. **Mayor, H. D. and Jordan, L. E.,** Electron microscopic study of the rodent "picodnavirus" X14, *Exp. Mol. Pathol.,* 5, 580, 1966.
9. **Mayor, H. D. and Melnick, J. L.,** Small deoxyribonucleic acid-containing viruses (picodnavirus group), *Nature (London),* 210, 331, 1966.
10. **Rose, J. A.,** Parvovirus reproduction, in *Comprehensive Virology,* Vol. 3, Fraenkel-Conrat, H. and Wagner, R. R., Eds., Plenum Press, New York, 1974, 1.
11. **Smith, K. O., Gehle, W. D., and Thiel, J. T.,** Properties of a small virus associated with adenovirus type 4, *J. Immunol.,* 97, 754, 1966.
12. **Tinsley, T. W. and Layworth, J. F.,** Parvoviruses, *J. Gen. Virol.,* 20, 71, 1973.
13. **Toolan, H. W., Saunders, E. L., Green, E. L., and Fabrizio, D. P. A.,** Further studies on the electron microscopy of the H-1 virus, *Virology,* 22, 286, 1964.
14. **Vasquez, C. and Brailovsky, C.,** Purification and fine structure of Kilham's rat virus, *Exp. Mol. Pathol.,* 4, 130, 1965.

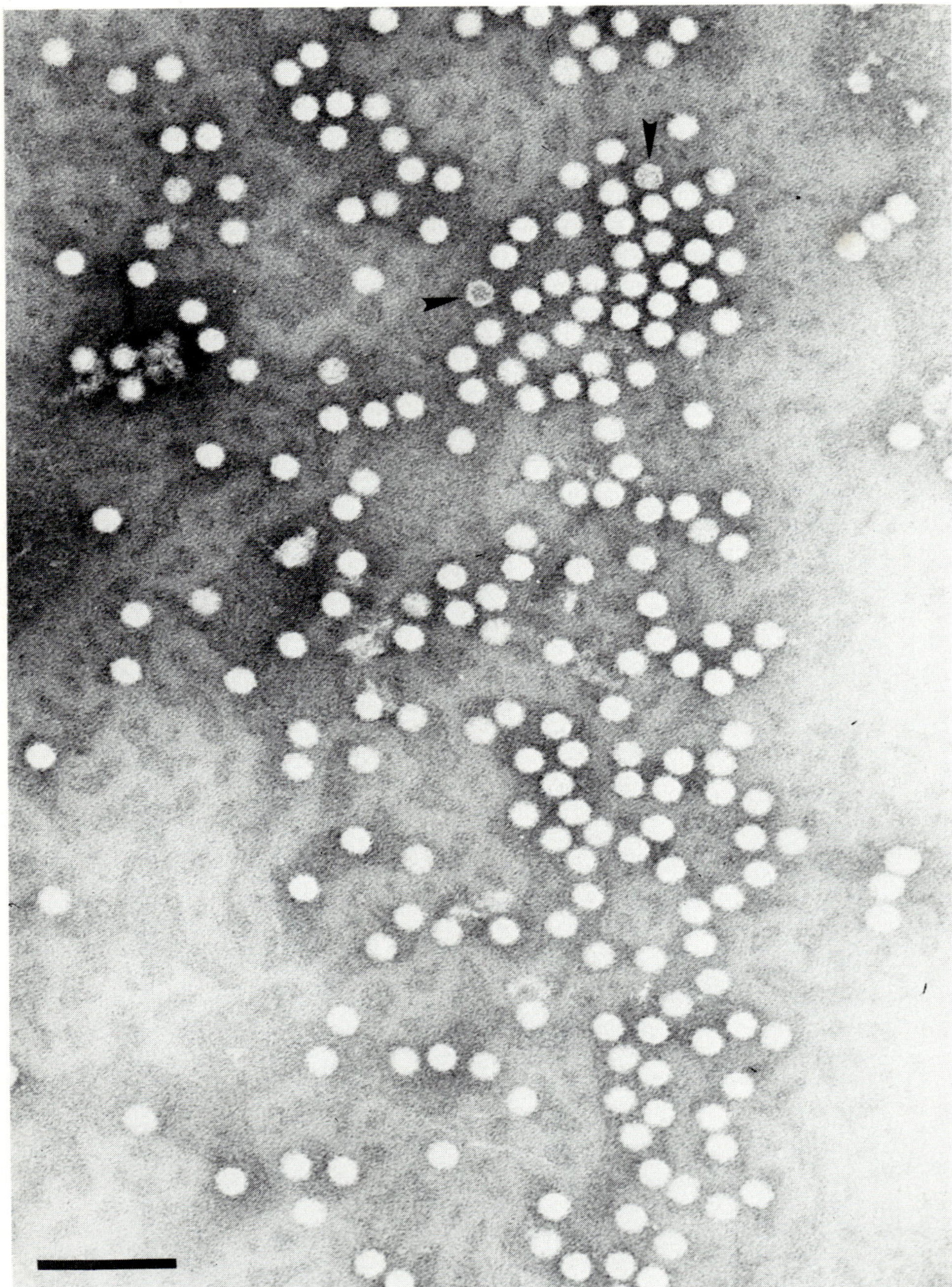

PLATE 1.   Parvovirus H-1. Some particles are penetrated by stain (arrows) and show an isometric configuration. Capsomere arrangement cannot be discerned because of the small size of these viruses. They usually cannot be distinguished from ribosomes of disrupted cells unless seen in groups of three or more. All parvoviruses have the same morphology and do not vary much in size. All bars equal 100 nm.

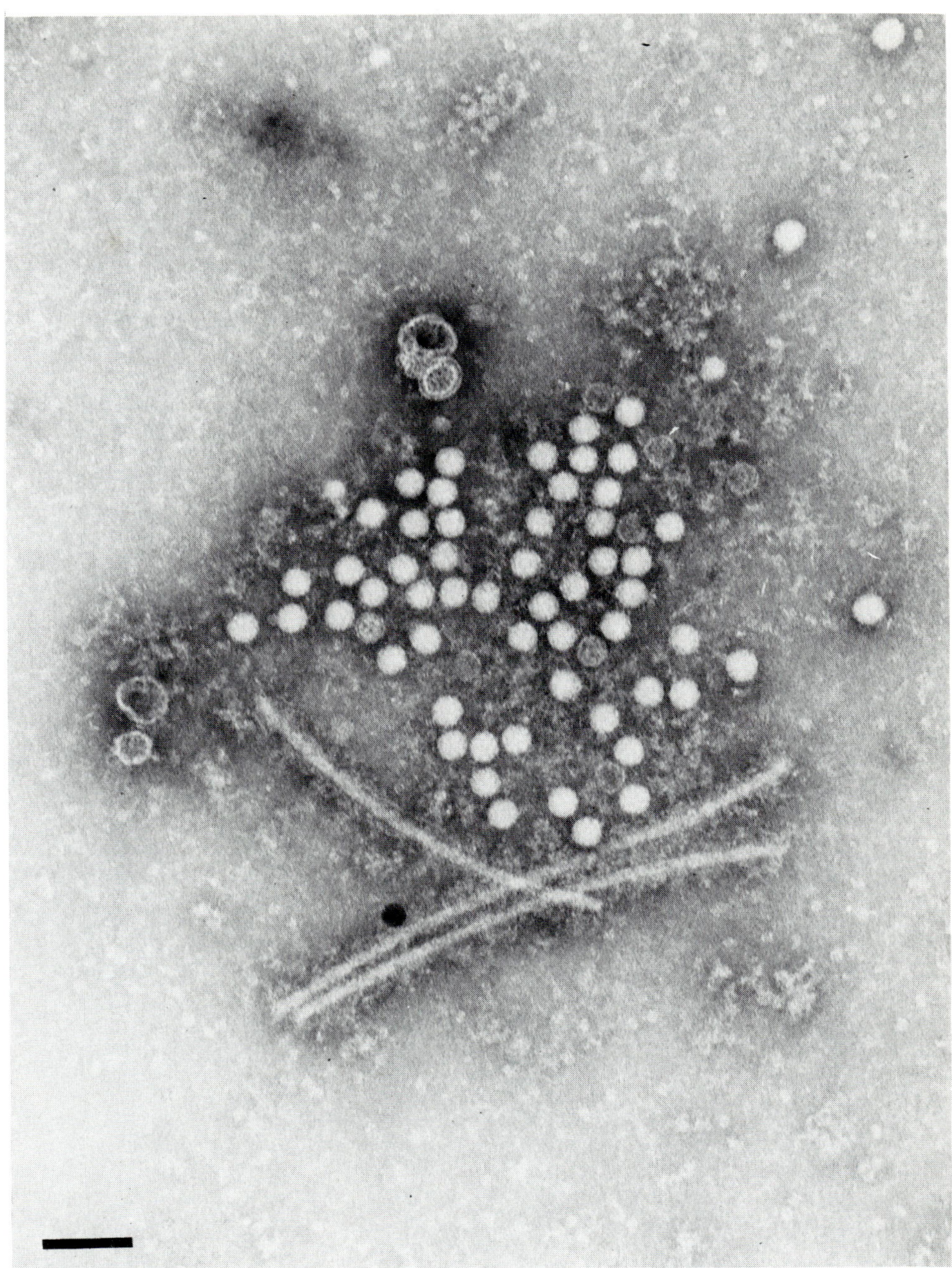

PLATE 2.   Group of parvo-like particles in the stool of a person with gastroenteritis.

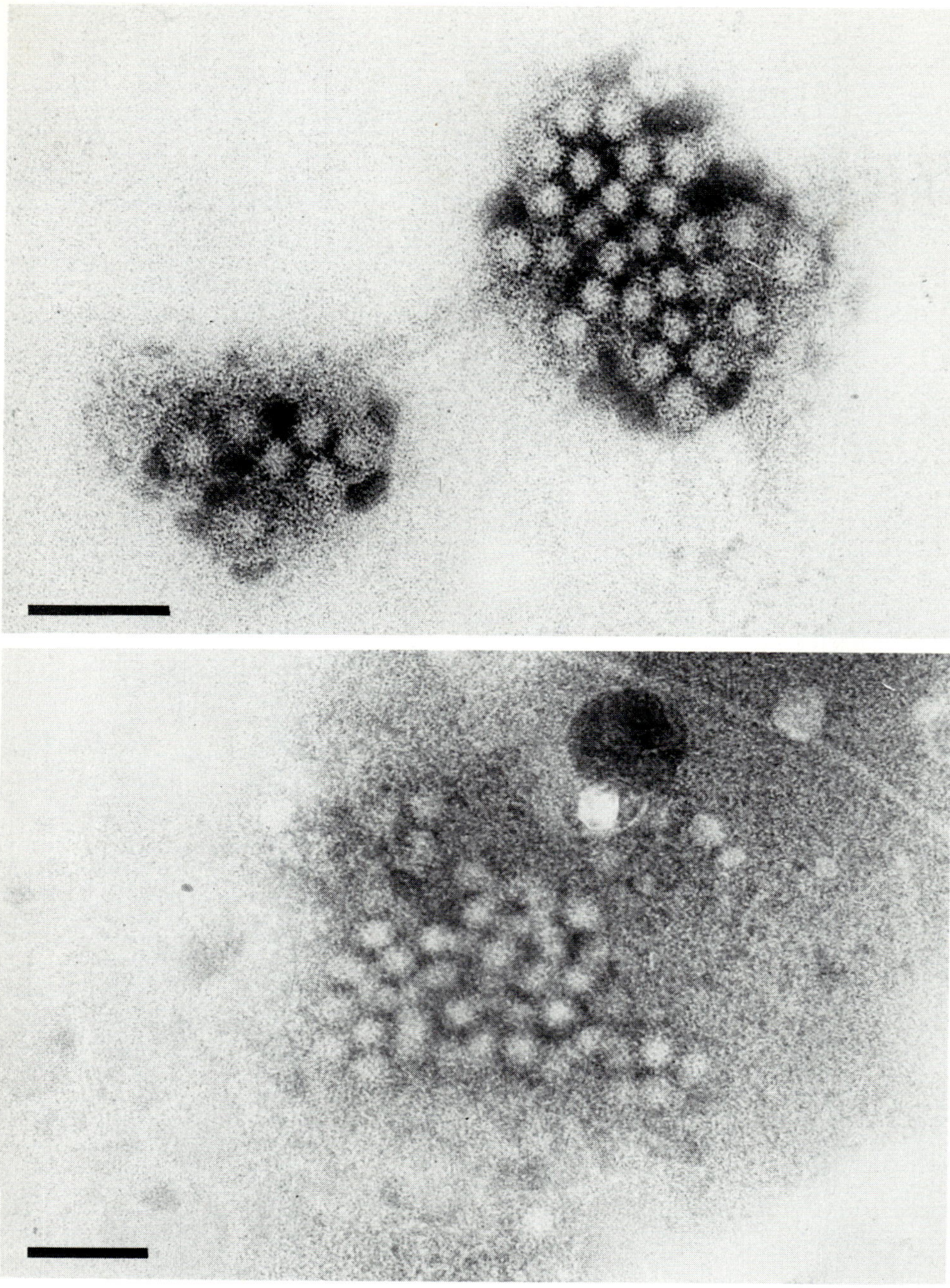

PLATE 3.   Immune electron microscopy of fecal extracts with hyperimmune serum to the Norwalk agent. Particles are not very clear because of antibody on the surface.

Chapter 16

PAPOVAVIRIDAE

The prefix "papova-" in papovavirus is derived from *papilloma*, *polyoma*, and *vacuolating agent* (SV40). The family consists of two genera, *Papillomavirus* and *Polyomavirus*. Viruses of the papillomavirus (from the Latin papilla for nipple or pustule, and the Greek suffix oma to form nouns denoting tumorous) group produce benign warts in a variety of mammals including man. The type species is Shope rabbit papilloma. The papillomas of rabbits may progress to neoplasia after several months. The papilloma viruses do not replicate in tissue culture but are readily seen by EM in extracts of human warts. Virions are 45 to 55 nm in diameter. They are composed of 72 capsomeres in a skew arrangement with T = 7 symmetry. Human papilloma virus has a right-handed, or dextro, arrangement of capsomeres, whereas rabbit papilloma virus has a left-handed, or levo, capsomere arrangement. The DNA is double-stranded within the naked nucleocapsid.

The genus *Polyomavirus* contains the polyomavirus of mice, BK and JC viruses of man, RK virus of rabbits, and SV 40 of monkeys. Polyomaviruses are slightly smaller than viruses of the *Papillomavirus* genus. The diameter of the capsid is 45 nm. The DNA is double-stranded and cyclic and is infectious when separated from the capsid. There are 72 capsomeres in a skew arrangement forming the naked nucleocapsid.

Polyomaviruses cause tumors of many types when inoculated into a variety of laboratory animals. SV40 causes productive infection of African green monkey kidneys and is a frequent contaminant of rhesus monkey kidney cell cultures. BK virus is sometimes found in the urine of organ transplant patients, and JC virus has been recovered from the brains of persons who have died of progressive multifocal leukoencephalopathy.

REFERENCES

1. **Almeida, J. A., Howatson, A. F., and Williams, M. G.,** Electron microscope study of human warts, sites of virus production and nature of the inclusion bodies, *J. Invest. Dermatol.,* 38, 337, 1962.
2. **Anderer, F. A., Schlumberger, H. D., Koch, M. A., Frank, H., and Eggers, H. J.,** Structure of simian virus 40. II. Symmetry and components of the virus particle, *Virology,* 32, 511, 1967.
3. **Crowther, R. A. and Amos, L. A.,** Three-dimensional image reconstruction of some small spherical viruses, *Cold Spring Harbor Symp. Quant. Biol.,* 36, 489, 1971.
4. **Finch, J. T.,** The surface structure of polyoma virus, *J. Gen. Virol.,* 24, 359, 1974.
5. **Finch, J. T. and Klug, A.,** The structure of viruses of the papilloma-polyoma type. III. Structure of rabbit papilloma virus, *J. Mol. Biol.,* 13, 1, 1965.
6. **Finch, J. T. and Crawford, L. V.,** Structure of small DNA-containing animal viruses, in *Comprehensive Virology,* Vol. 5, Fraenkel-Conrat, H. and Wagner, R. R., Eds., Plenum Press, New York, 1975, 119.
7. **Howatson, A. F. and Crawford, L. V.,** Direct counting of the capsomers in polyoma and papilloma viruses, *Virology,* 21, 1, 1963.
8. **Kiselev, N. A. and Klug, A.,** The structure of viruses of the papilloma-polyoma type. V. Tubular variants built of pentamers, *J. Mol. Biol.,* 40, 155, 1969.
9. **Klug, A.,** The structure of viruses of the papilloma-polyoma type. II. Comments on other work, *J. Mol. Biol.,* 11, 424, 1965.
10. **Klug, A. and Finch, J. T.,** The structure of viruses of the papilloma-polyoma type. I. Human wart virus, *J. Mol. Biol.,* 11, 403, 1965.
11. **Klug, A. and Finch, J. T.,** The structure of viruses of the papilloma-polyoma type. IV. Tilting experiments and two side images, *J. Mol. Biol.,* 31, 1, 1968.
12. **Melnick, J. L., Allison, A. C., Butel, J. S., Eckhart, W., Eddy, B. E., Kit, S., Levine, A. J., Miles, J. A. R., Pagano, J. S., Sachs, L., and Vonka, V.,** Papovaviridae, *Intervirology,* 3, 106, 1974.
13. **Shope, R. E.,** Infectious papillomatosis of rabbits, *J. Exp. Med.,* 58, 607, 1933.
14. **Takemoto, K. K., Mattern, C. F. T., and Murakami, W. T.,** The papovavirus group, in *Comparative Virology,* Maramorosch, K. and Kurstak, E., Eds., Academic Press, New York, 1971, 81.

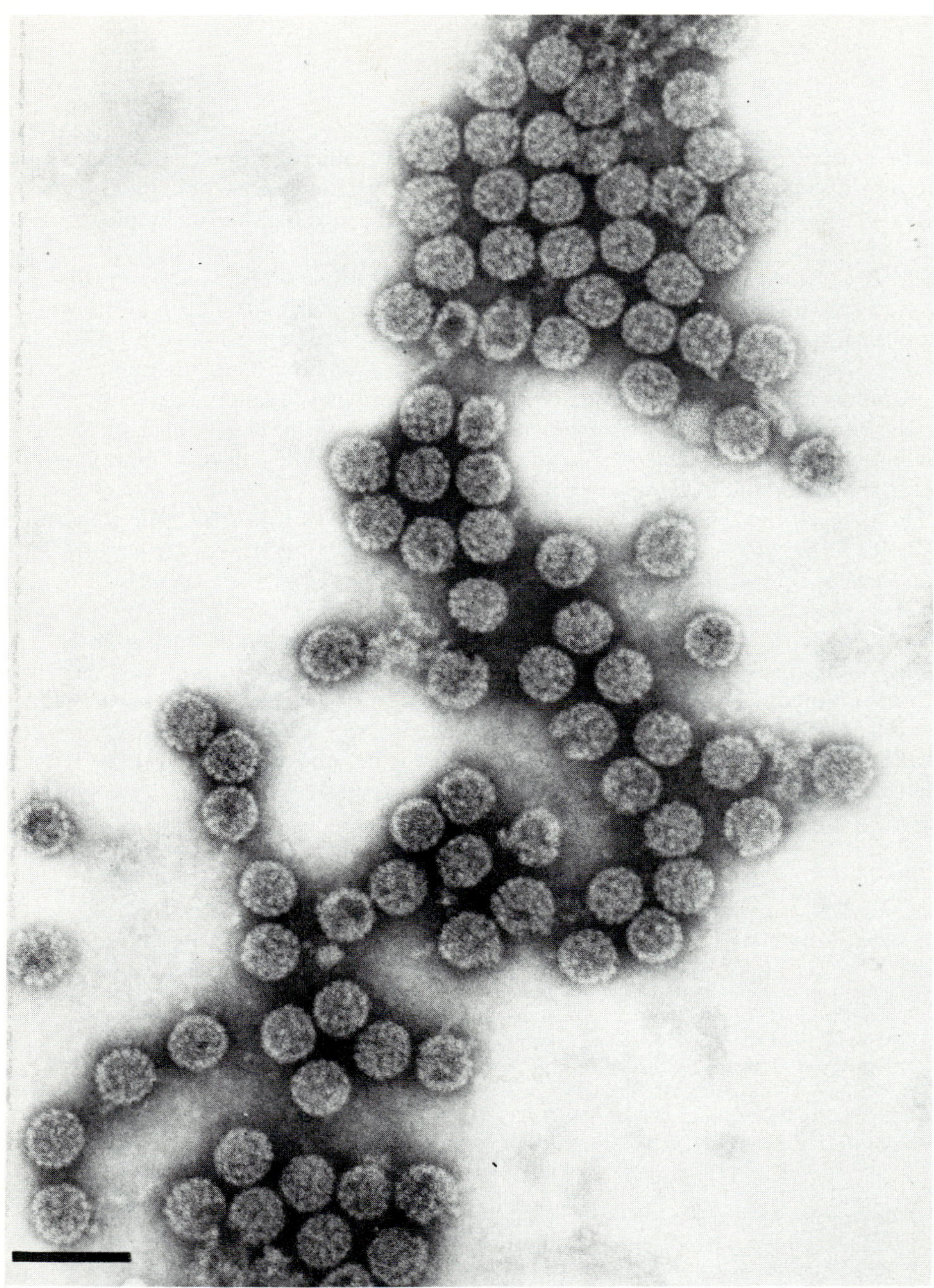

PLATE 1.   Survey micrograph of the papovavirus SV40. All bars equal 100 nm.

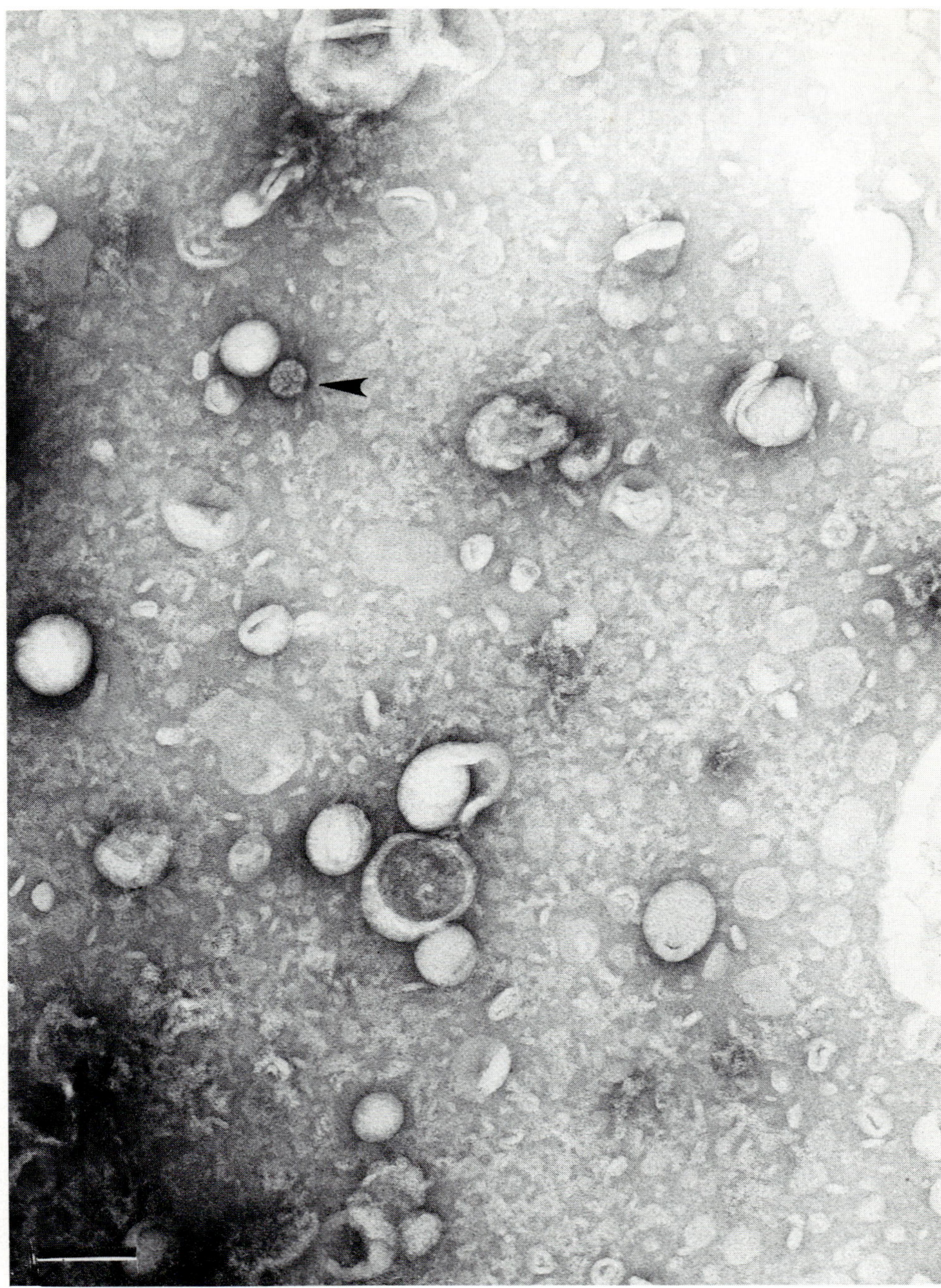

PLATE 2.   Polyomavirus JC (arrow) among debris of cells from a brain biopsy of a person who had progressive multifocal leukoencephalopathy.

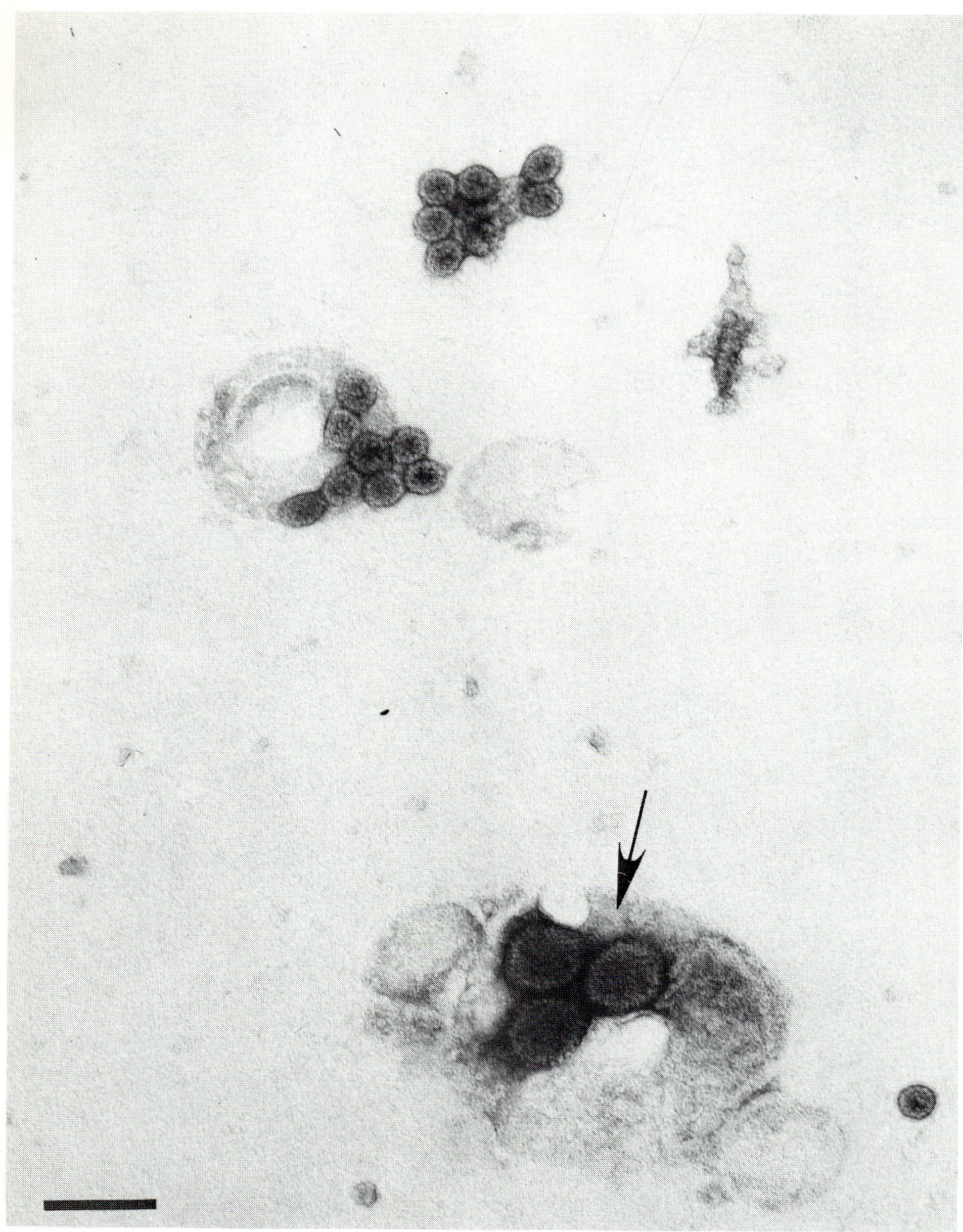

PLATE 3.   SV40 and adenovirus (arrow) from monkey kidney cells.

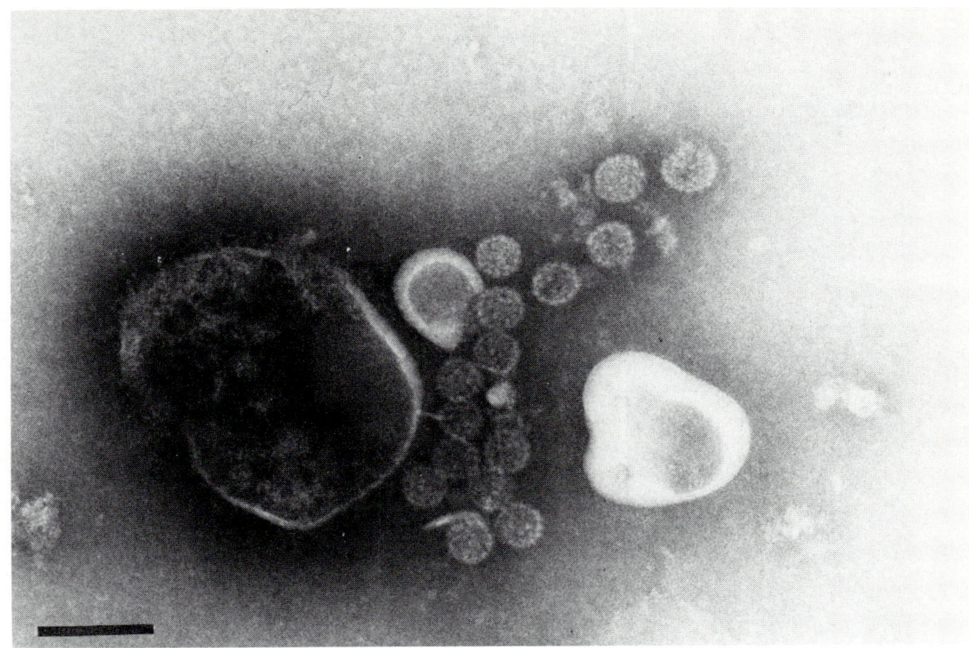

A

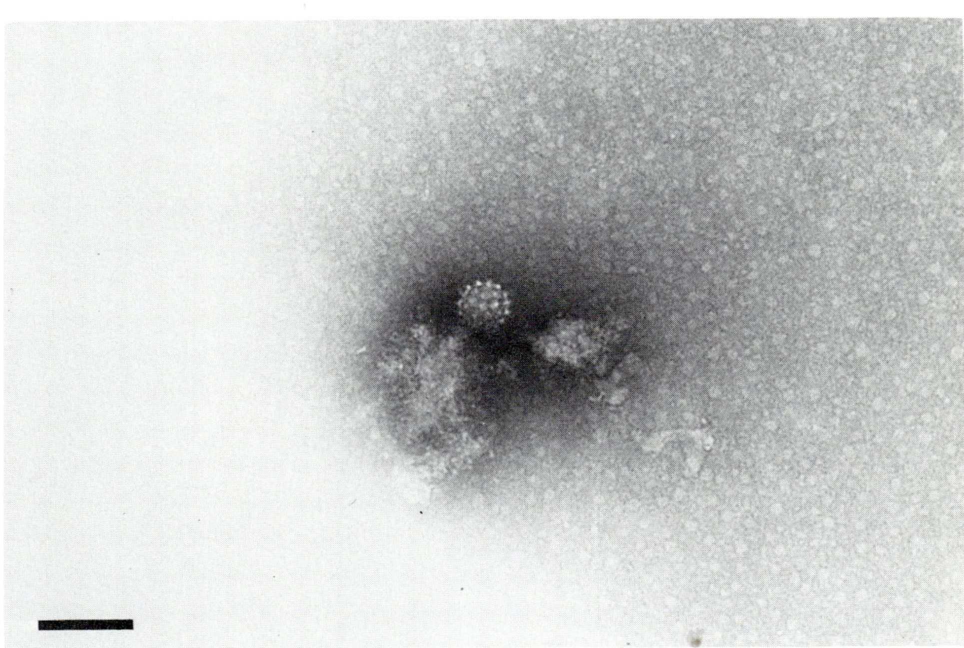

B

PLATE 4.   (A) Papovavirus from monkey tissue culture. (B) Papilloma virus from human wart material. The structure of papilloma viruses is usually better defined than that of polyoma viruses derived from tissue culture.

Chapter 17

ADENOVIRIDAE

Viruses of the Adenoviridae (aden, adenos = gland) were first isolated from human adenoid tissue in 1953. They have since been shown to be very wide spread in nature. At least 31 types affect humans and some 46 others are known to affect other mammals. The mammalian species share a common antigen and have been placed in the genus *Mastadenovirus*. There are also several species of avian adenoviruses (genus *Aviadenovirus*).

The mammalian adenoviruses induce latent infections in lymphoid tissue and are readily activated. Type species may cause such maladies as acute respiratory disease of military recruits, conjunctivitis, keratoconjunctivitis, and pharyngoconjunctival fever. Some strains produce tumors when inoculated into newborn rats, hamsters and mice. Adenoviruses have also been found in association with gastroenteritis in humans. These latter viruses may constitute a new serotype.

Morphologically, the adenovirus particle is 70 to 90 nm in diameter. It is a DNA-containing naked icosahedron composed of 252 capsomeres which are 7 to 9 nm in diameter. There are 240 prismatic hexons which have group- and type-specific determinants. There are 12 vertex pentons from which fibers or filaments project. The fibers contain major type-specific and minor subgroup antigens. Tips of the penton fibers bind to red blood cells and cause cross-linking and agglutination. The mammalian adenoviruses can be roughly grouped into three groups on the basis of ability to agglutinate rhesus monkey or rat erythrocytes.

REFERENCES

1. **Crowther, R. A. and Franklin, R. M.,** The structure of the groups of nine hexons from adenovirus, *J. Mol. Biol.,* 68, 181, 1972.
2. **Everitt, E., Lutter, L., and Philipson, L.,** Structural proteins of adenovirus, *Virology,* 67, 197, 1975.
3. **Ginsberg, H. S.,** Adenovirus structural proteins, in *Comprehensive Virology,* Vol. 13, Fraenkel-Conrat, H. and Wagner, R. R., Eds., Plenum Press, New York, 1979, 409.
4. **Ginsberg, H. S., Pereira, H. G., Valentine, R. C., and Wilcox, W. C.,** A proposed terminology for the adenovirus antigens and virion morphological subunits, *Virology,* 28, 782, 1966.
5. **Hilleman, M. R. and Werner, J. R.,** Recovery of new agent from patients with acute respiratory illness, *Proc. Soc. Exp. Biol. Med.,* 85, 183, 1954.
6. **Horne, R. W.,** The comparative structure of adenoviruses, *Ann. N. Y. Acad. Sci.,* 101, 475, 1962.
7. **Horne, R. W., Brenner, S., Waterson, A. P., and Wildy, P.,** The icosahedral form of an adenovirus, *J. Mol. Biol.,* 1, 84, 1959.
8. **Horne, R. W., Ronchetti, I. P., and Hobart, J. M.,** A negative staining-carbon film technique for studying viruses in the electron microscope. II. Application to adenovirus type 5, *J. Ultrastruct. Res.,* 51, 233, 1975.
9. **Nermut, M. V.,** Fine structure of adenovirus type 5, *Virology,* 65, 480, 1975.
10. **Norrby, E.,** The structural and functional diversity of adenovirus capsid components, *J. Gen. Virol.,* 5, 221, 1969.
11. **Pereira, H. G. and Wrigley, N. G.,** *In vitro* reconstitution, hexon bonding and handedness of incomplete adenovirus capsid, *J. Mol. Biol.,* 85, 617, 1974.
12. **Pettersson, U. and Hoglund, S.,** Structural proteins of adenoviruses. III. Purification and characterization of adenovirus type 2 penton antigen, *Virology,* 39, 90, 1969.
13. **Philipson, L. and Lindberg, U.,** Reproduction of adenoviruses, in *Comprehensive Virology,* Vol. 3, Fraenkel-Conrat, H. and Wagner, R. R., Eds., Plenum Press, New York, 1974, 143.
14. **Philipson, L., Pettersson, U., and Lindberg, U.,** *Molecular Biology of Adenoviruses, Virology Monogr.,* 14, 1, 1975.

15. **Prage, L., Pettersson, U., Hoglund, S., Lenberg-Holm, K., and Philipson, L.,** Structural proteins of adenoviruses. IV. Sequential degradation of the adenovirus type 2 virion, *Virology,* 42, 341, 1970.
16. **Smith, K. O., Gehle, W. D., and Trousdale, M. D.,** Architecture of the adenovirus capsid, *J. Bacteriol.,* 90, 254, 1965.
17. **Valentine, R. C. and Pereira, H. G.,** Antigens and structure of the adenovirus, *J. Mol. Biol.,* 13, 13, 1965.
18. **Wilcox, W. C., Ginsberg, H. S., and Anderson, T. T.,** Structure of type 5 adenovirus. II. Fine structure of virus subunits. Morphologic relationship of structural subunits to virus-specific antigen from infected cells, *J. Exp. Med.,* 118, 307, 1963.

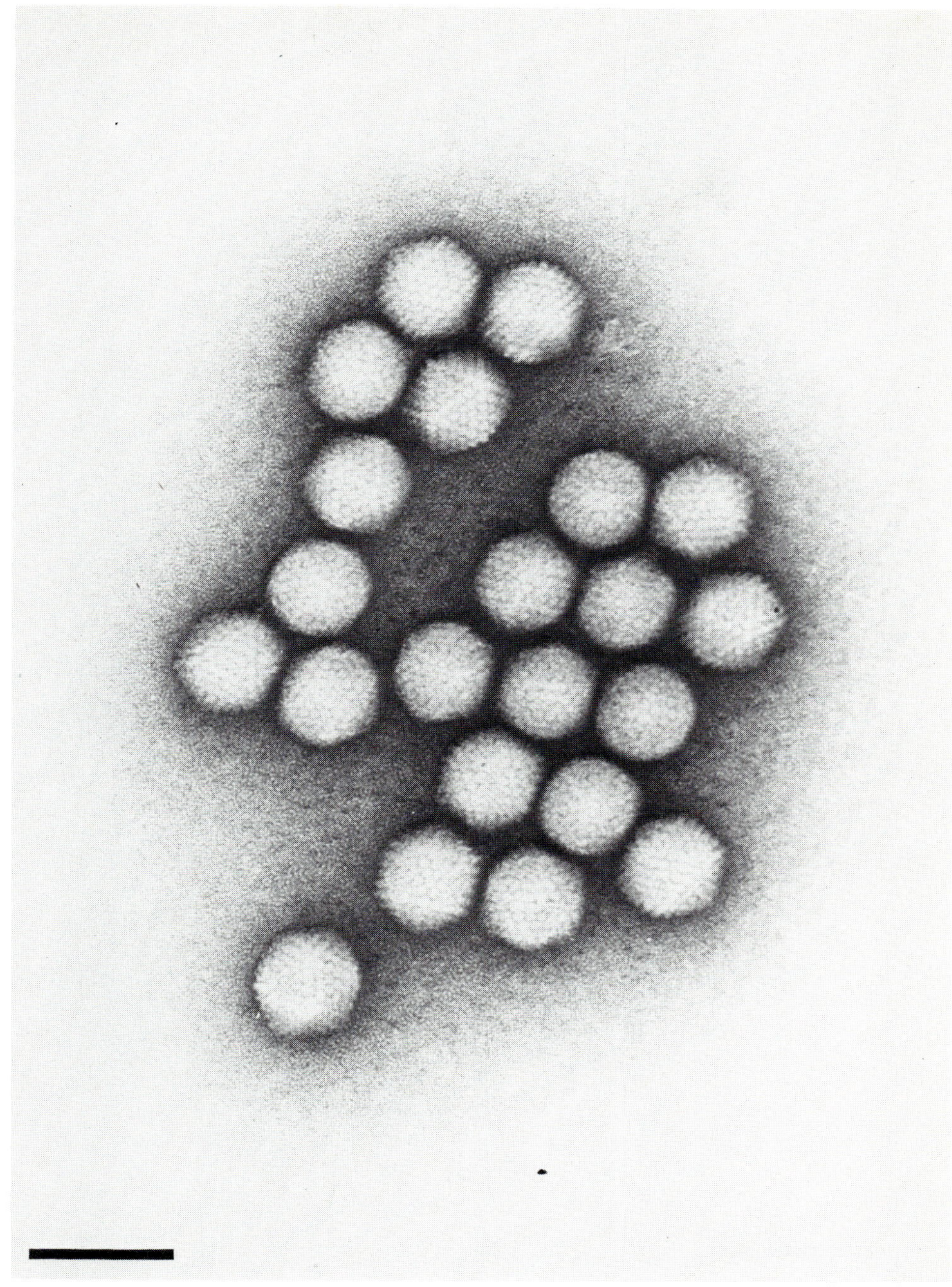

PLATE 1. Group of adenovirus particles isolated from a stool extract of a child with gastroenteritis. All adenoviruses have identical morphology. All bars equal 100 nm.

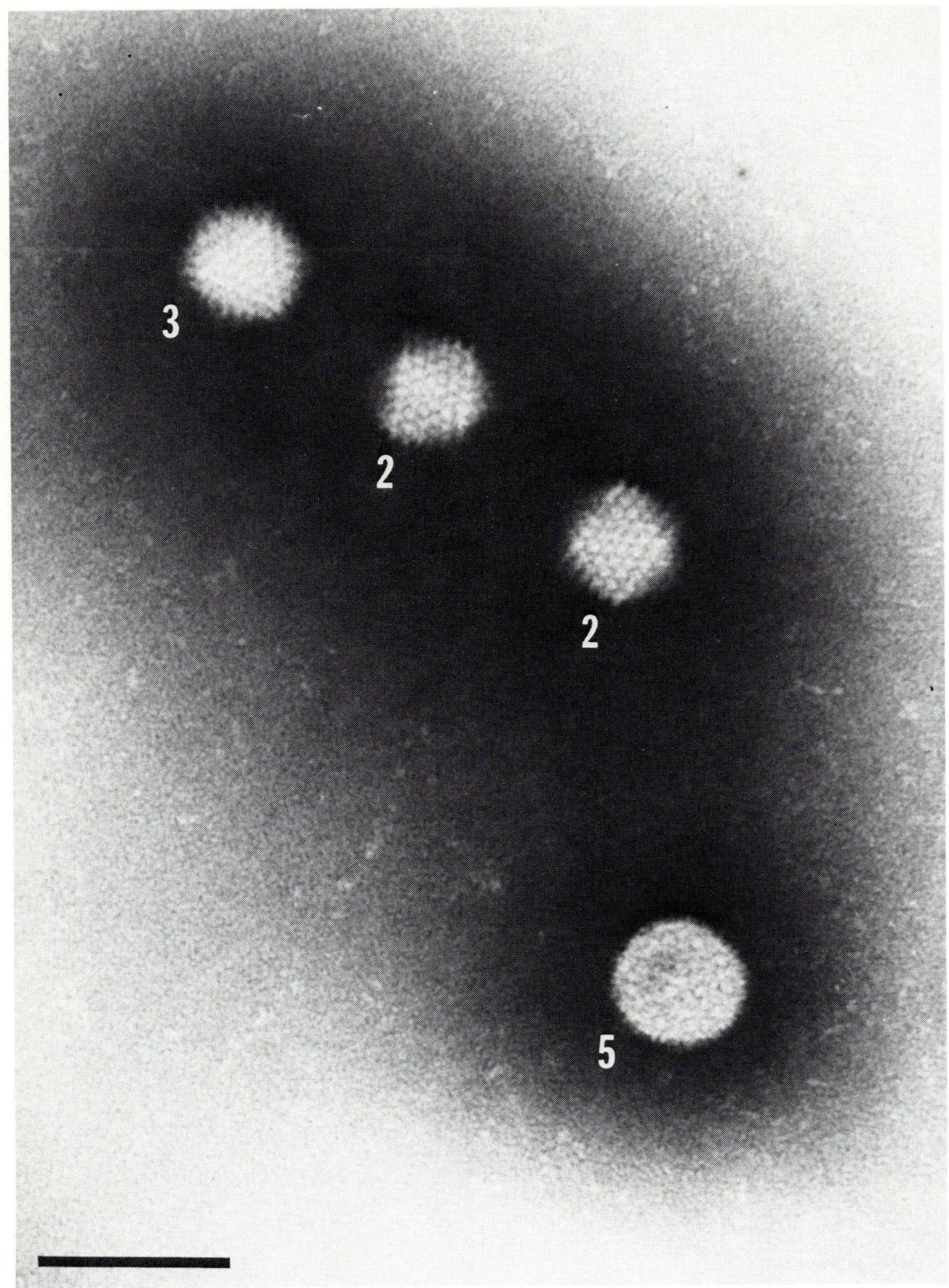

PLATE 2.   Adenovirus particles seen on axes of three, two, and fivefold symmetry.

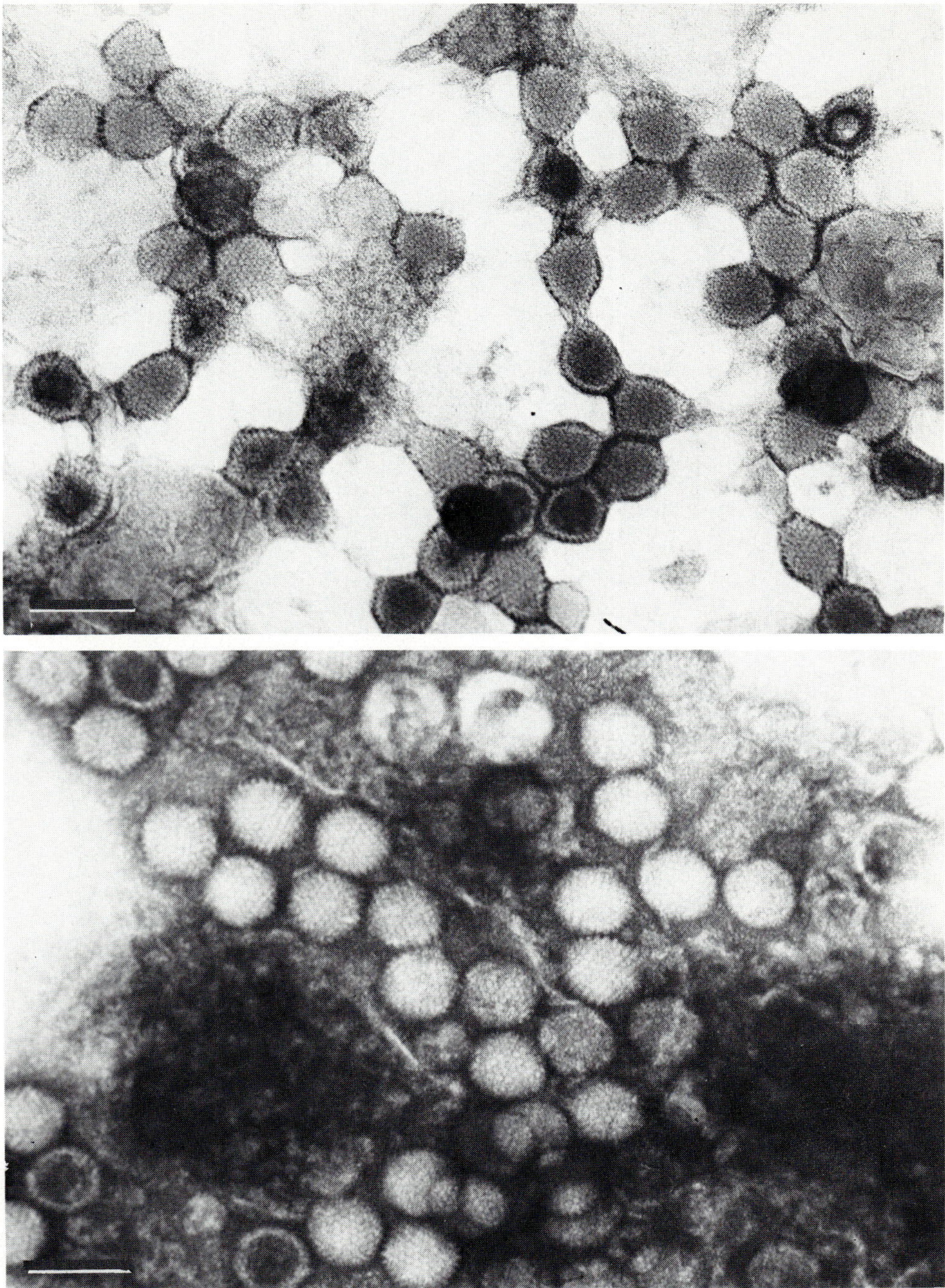

PLATE 3.   Adenovirus from tissue cultures inoculated with nasal and throat washings.

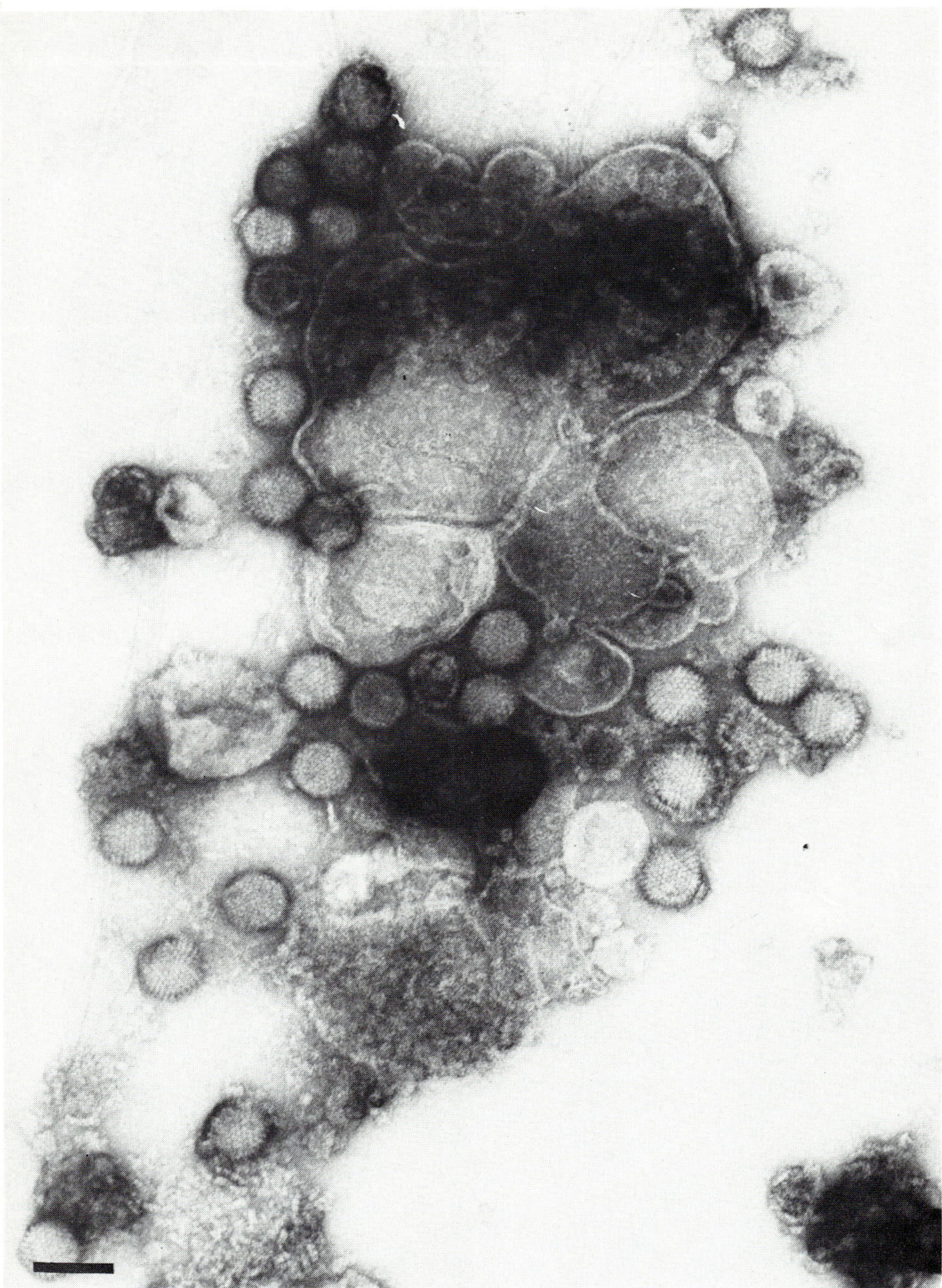

PLATE 4.   Adenovirus in Hep 2 cells inoculated with a throat washing. Large blebs are cell membrane debris.

*Part VI
DNA Viruses with
Cubic Capsid Symmetry and
Complex Envelopes*

Chapter 18

HERPESVIRIDAE

Herpes simplex virus (herpein = to creep) was the first animal virus to be thoroughly examined by negative contrast electron microscopy. Wildy et al.,[20] in their classical description of this virus, demonstrated that it consisted of a nucleocapsid surrounded by an envelope, and from the architecture of the capsid they deduced that it was composed of 162 capsomeres. Since then viruses with similar morphology have been isolated from numerous animal species. Those important to humans include herpes simplex virus type 1 and type 2, varicella-zoster virus, cytomegalovirus, Epstein-Barr virus, and herpes simiae (B virus). Herpes simplex and herpes simiae viruses are closely related serologically, and there is some evidence that herpes simplex and varicella-zoster viruses may share a minor precipitin antigen.

Wildy et al.,[20] described two basic types of herpesvirus particles: one with a naked nucleocapsid and the other with a nucleocapsid surrounded by an envelope. They were able to determine that the capsid was composed of capsomeres arrayed as an icosahedron, with 5 capsomeres along each edge of the triangular facets of the icosahedron so that each capsid had a total of 162 capsomeres $[10(n - 1)^2 + 2$, where n = 5]. The nucleocapsid is seen either as hexagonal or spherical and measures 100 to 105 nm in diameter. The capsomeres comprising the capsid are hollow prisms. Measurements of the prisms from angle to angle range from 11 to 12 nm, and measurements from opposing flat sides are approximately 9.5 nm. The axial hole is 4 to 5 nm in diameter. Intercapsomeric linkages which originate from the prism edges join capsomeres together. One tripartite strand links adjacent capsomeres and the central capsomere of a hexagonal or pentagonal cluster.

The nucleocapsid forms in the nucleus of infected cells, and most herpes viruses derive their envelopes from the inner nuclear membrane or other cellular membranes. Enveloped particles usually contain only one nucleocapsid, but it is not uncommon to see two or three inside one envelope. Enveloped particles vary greatly in size when seen in negative stain preparations but are usually 150 to 180 nm. When enveloped particles are not penetrated by stain they are difficult to distinguish from cell membrane material. Enveloped particles also have regularly spaced surface projections which are about 10 nm in length. The significance of these projections is not well understood.

In micrographs of thin sections, a herpesvirus core measuring about 30 nm can be resolved. It encloses a double-stranded DNA genome. This structure is not readily isolated, and little is known about its morphology. There is also some evidence that herpesviruses have a middle capsid layer just beneath the capsid and an inner capsid about 45 nm in diameter which encloses the core. However, substantiation of such discrete structures is not yet unequivocal. For resolution of these components as discrete structures, a gentle method of controlled degradation of herpes virions must be found.

REFERENCES

1. **Almeida, J. D., Howatson, A. F., and Williams, M.,** Morphology of varicella (chickenpox) virus, *Virology,* 16, 353, 1962.
2. **Cook, M. L. and Stevens, J. G.,** Replication of varicella-zoster virus in cell cultures: an ultrastructural study, *J. Ultrastruct. Res.* 32, 334, 1970.
3. **Darlington, R. W. and Moss, L. H., III,** The envelope of herpes virus, *Progr. Med. Virol.,* 11, 16, 1969.

4. **Epstein, M. A. and Achong, B. G.,** Morphology of the virus and of virus-induced cytopathologic changes, in *The Epstein-Barr Virus,* Epstein, M. A. and Achong, B. G., Eds., Springer-Verlag, Berlin, 1979, 23.

5. **Furlong, D., Swift, H., and Roizman, B.,** Arrangement of herpesvirus deoxyribonucleic acid in the core, *J. Virol.,* 10, 1071, 1972.

6. **Horne, R. W. and Wildy, P.,** Symmetry in virus architecture, *Virology,* 15, 348, 1961.

7. **Morgan, C., Ellison, S. A., Rose, H. M., and Moore, D. H.,** Structure and development of viruses as observed in the electron microscope. I. Herpes simplex virus, *J. Exp. Med.,* 100, 195, 1954.

8. **Nazerian, K.,** DNA configuration in the core of Marek's disease virus, *J. Virol.,* 13, 1148, 1974.

9. **Nii, S., Morgan, C., and Rose, H. M.,** Electron microscopy of herpes simplex virus. II. Sequence of development, *J. Virol.,* 2, 517, 1968.

10. **Nii, S. and Yasuda, I.,** Detection of viral cores having toroid structures in eight herpesviruses, *Biken J.,* 18, 41, 1975.

11. **Palmer, E. L., Martin, M. L., and Gary, G. W., Jr.,** The ultrastructure of disrupted herpesvirus nucleocapsids, *Virology,* 65, 260, 1975.

12. **Roizman, B., Spring, S. B., and Schwartz, J.,** The herpesvirion and its precursors made in productively and abortively infected cells, *Fed. Proc.,* 28, 1890, 1969.

13. **Roizman, B. and Spear, P. G.,** Herpesviruses: current information on the composition and structure, in *Comparative Virology,* Maramorosch, K. and Kurstak, E., Eds., Academic Press, New York, 1971, 135.

14. **Roizman, B. and Furlong, D.,** The replication of herpesviruses, *Comprehensive Virology,* Vol. 3, Fraenkel-Conrat, H. and Wagner, R. R., Eds., Plenum Press, New York, 1974, 229.

15. **Schwartz, J. and Roizman, B.,** Similarities and differences in the development of laboratory strains and freshly isolated strains of herpes simplex virus in HEp-2 cells: electron microscopy, *J. Virol.,* 4, 879, 1969.

16. **Spear, P. G. and Roizman, B.,** Proteins specified by herpes simplex virus. V. Purification and structural proteins of the virion, *J. Virol.,* 9, 143, 1972.

17. **Toplin, I. and Schidlovsky, G.,** Partial purification and electron microscopy of virus in the EB-3 cell line derived from a Burkitt lymphoma, *Science,* 152, 1084, 1966.

18. **Watson, D. H.,** The structure of animal viruses in relation to their biological functions, *Symp. Soc. Gen. Microbiol.,* 18, 207, 1968.

19. **Watson, D. H.,** Morphology, in *The Herpesviruses,* Kaplan, A. S., Ed., Academic Press, New York, 1973, 27.

20. **Wildy, P., Russell, W. C., and Horne, R. W.,** The morphology of herpes virus, *Virology,* 12, 204, 1960.

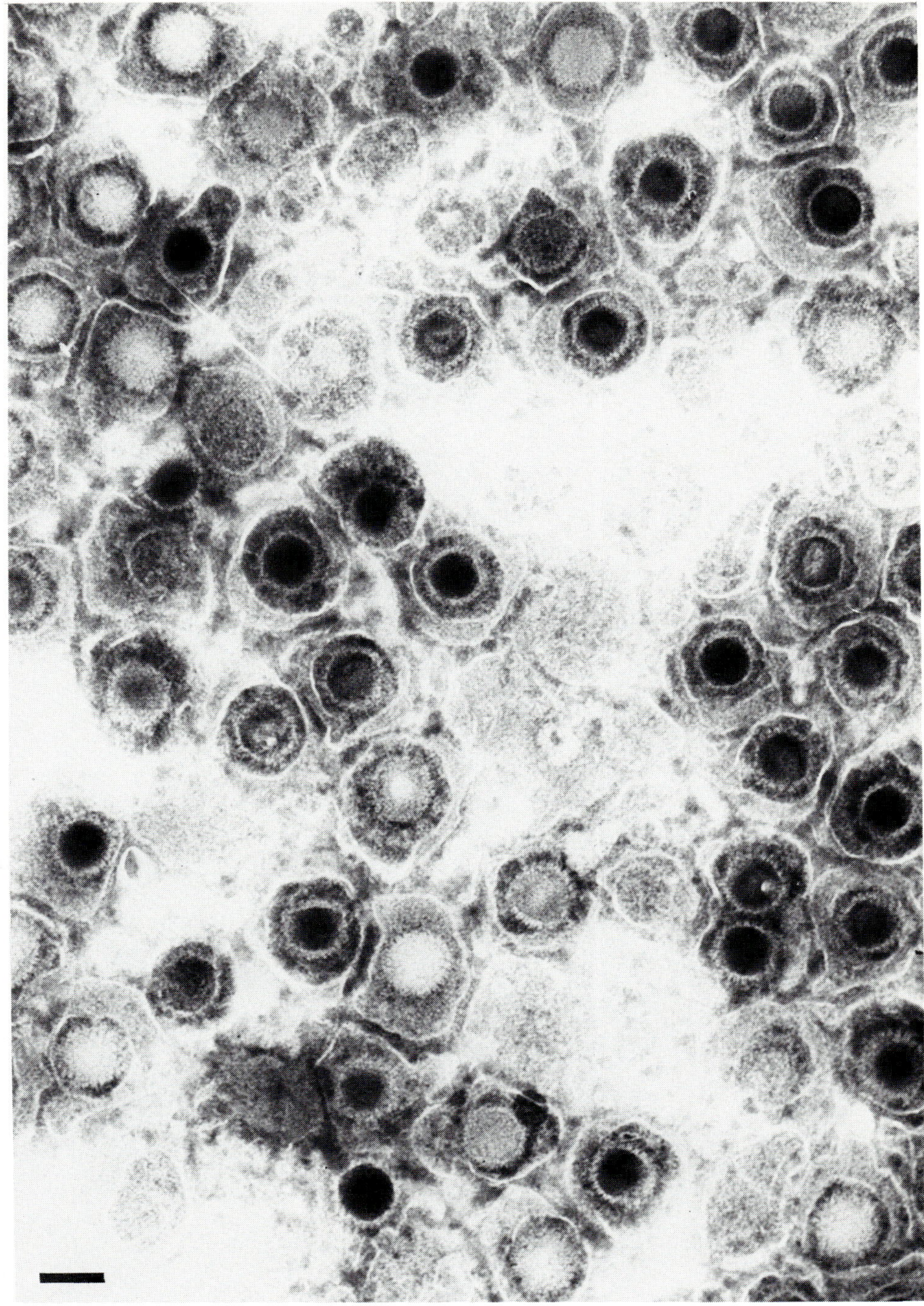

PLATE 1.   Group of herpesvirus particles. Some are penetrated by stain to reveal full (electron-lucent) and empty (electron-dense) nucleocapsids. The envelope of most particles is clearly visible and some particles have surface projections. All bars equal 100 nm except where indicated.

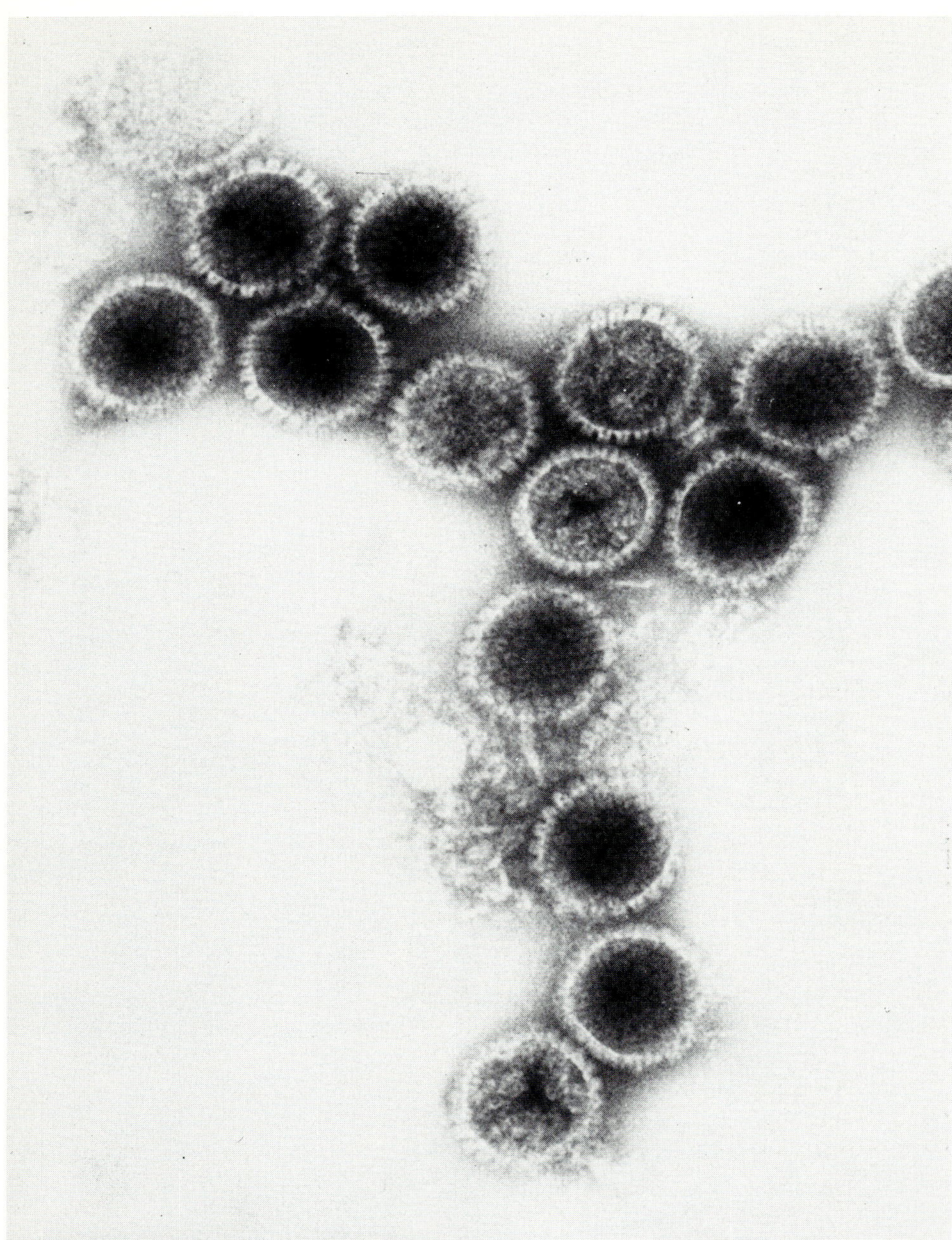

PLATE 2.   Group of herpesvirus nucleocapsids. Capsomeres are visible on the surface of most nucleo-capsids.

PLATE 3.   (A) Herpesvirus particle unpenetrated by stain. The "studded" appearance of the particle results because the viral envelope is not damaged and surface projections are viewed end-on. (B) Herpes-virus particle penetrated by stain to reveal "full" nucleocapsid. Capsomeres on the nucleocapsid can be seen in threefold symmetry. (C) Full herpesvirus particle showing inner (electron-dense) and outer enve-lope. (D) Full herpesvirus particle showing partial penetration of nucleocapsid to outline what may be the viral core. (E) Empty herpesvirus particle where stain has penetrated both the envelope and the nucleocapsid. A well-stained envelope layer and surface projections can also be seen (arrow). (F) Full herpesvirus nucleocapsid showing columnar prismatic capsomeres.

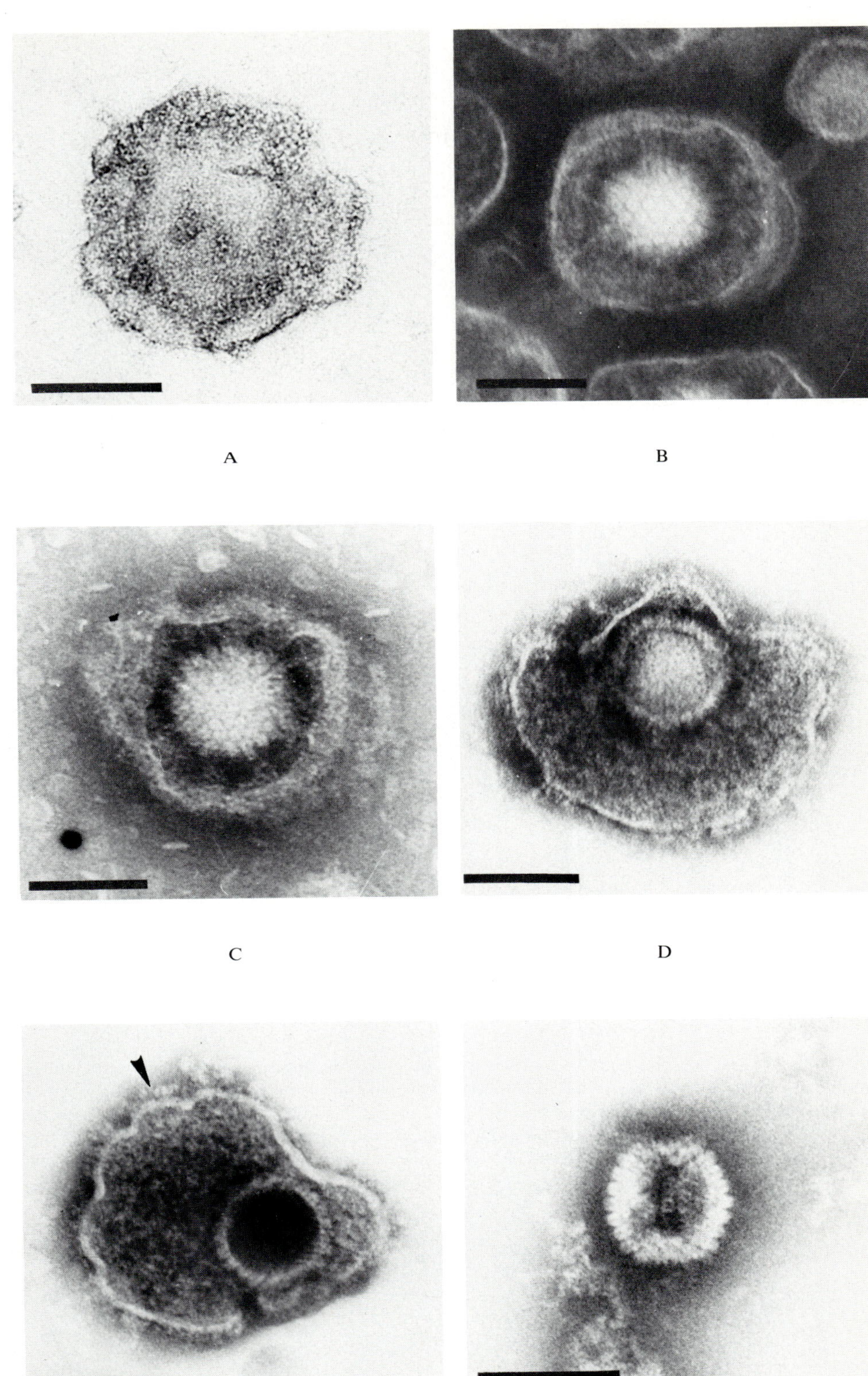

A

B

C

D

E

F

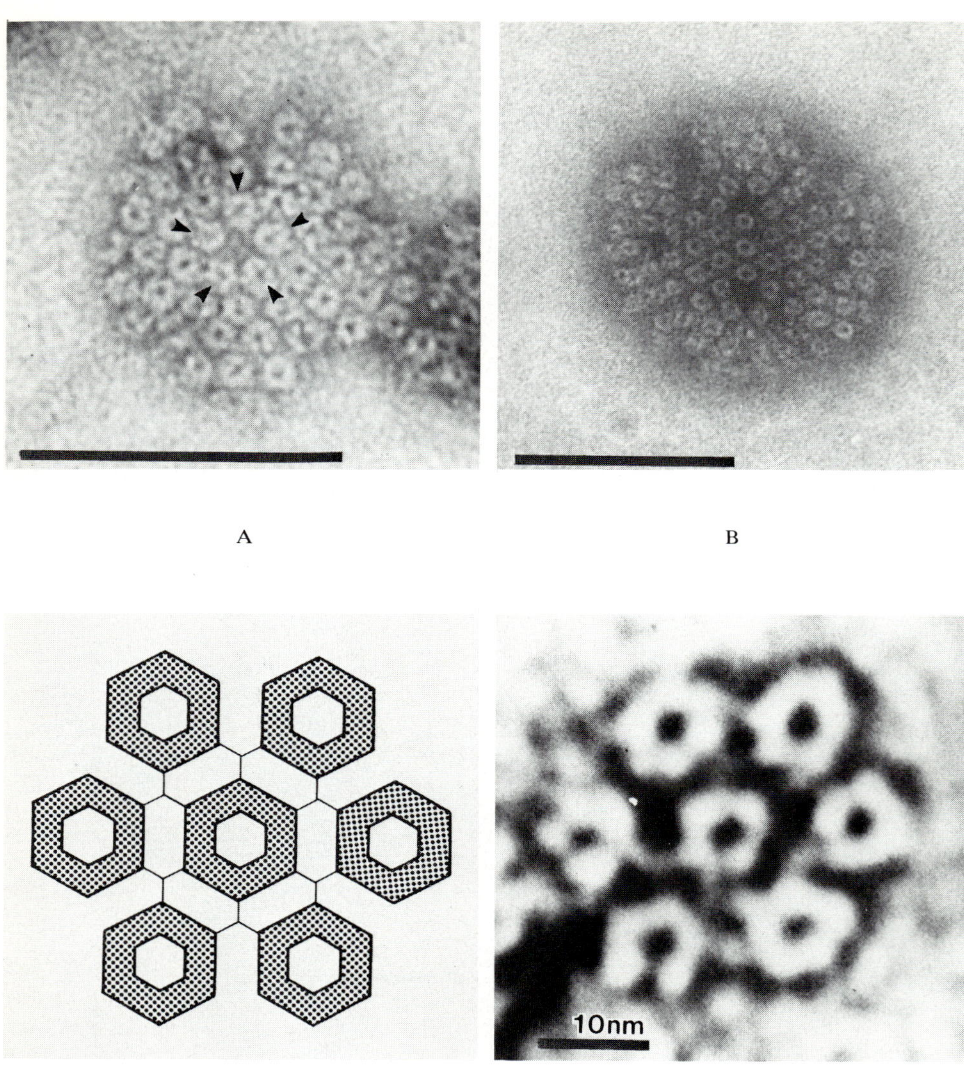

PLATE 4.   (A) Herpesvirus nucleocapsid which has been partially degraded by EDTA-trypsin showing capsomeres in fivefold symmetry (arrows). The pentagonal capsomere is missing. (B) Same as in 4a except particle is on an axis of twofold symmetry. The particles in Plate 4a and b also clearly show that the capsomeres are hollow prisms. (C) and (D) Diagram and micrograph of a group of six free-lying herpesvirus capsomeres from a preparation of virus disrupted with EDTA-trypsin. At this high magnification it can be seen that the capsomeres are joined by linkages as indicated in the diagram.

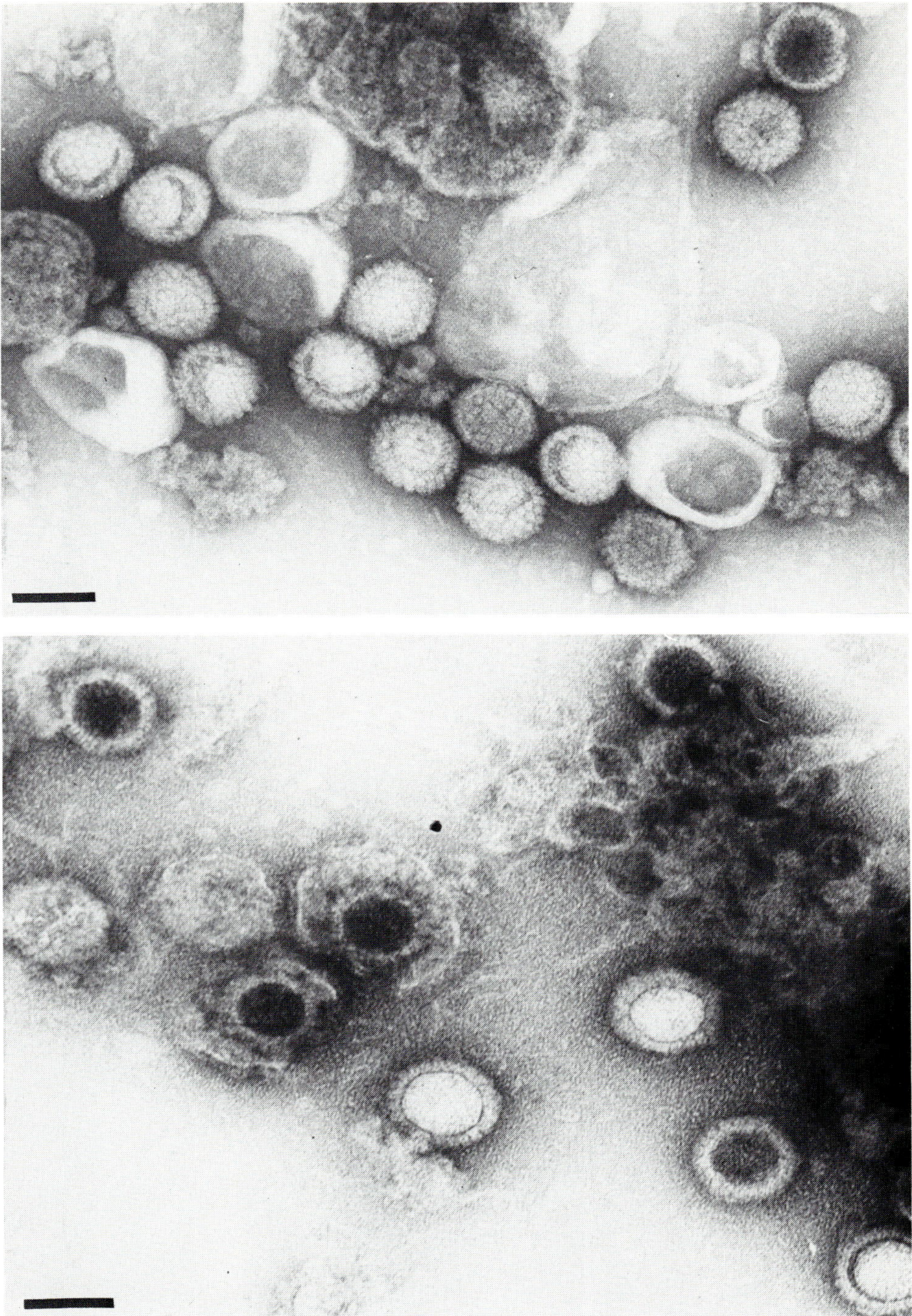

PLATE 5.   Herpesvirus in tissue culture. Nasal washings inoculated into tissue culture and culture fluids examined by EM. Top panel shows a group of nucleocapsids; bottom panel shows nucleocapsids, empty and enveloped particles, and unpenetrated particles.

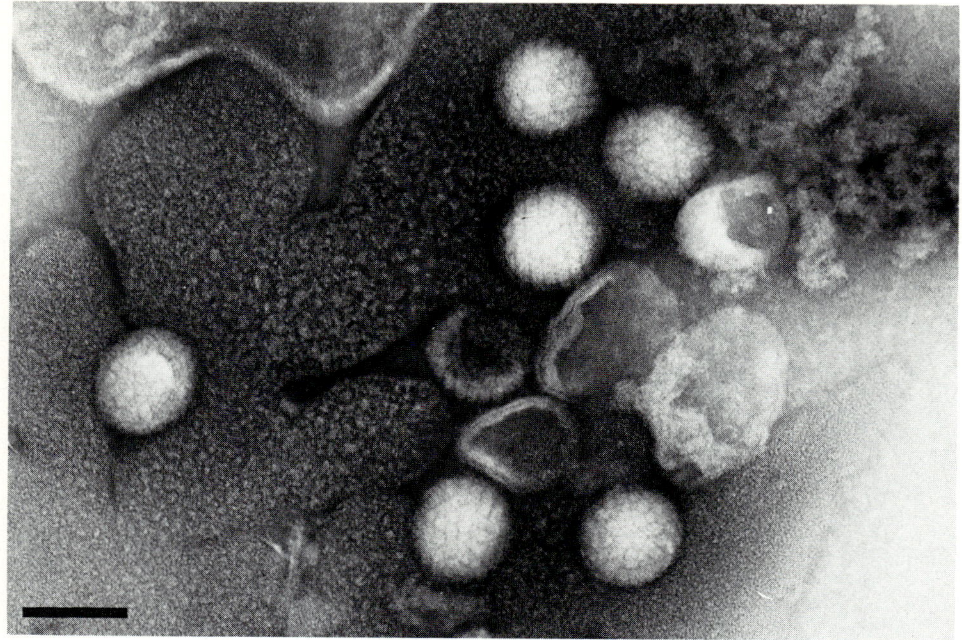

A

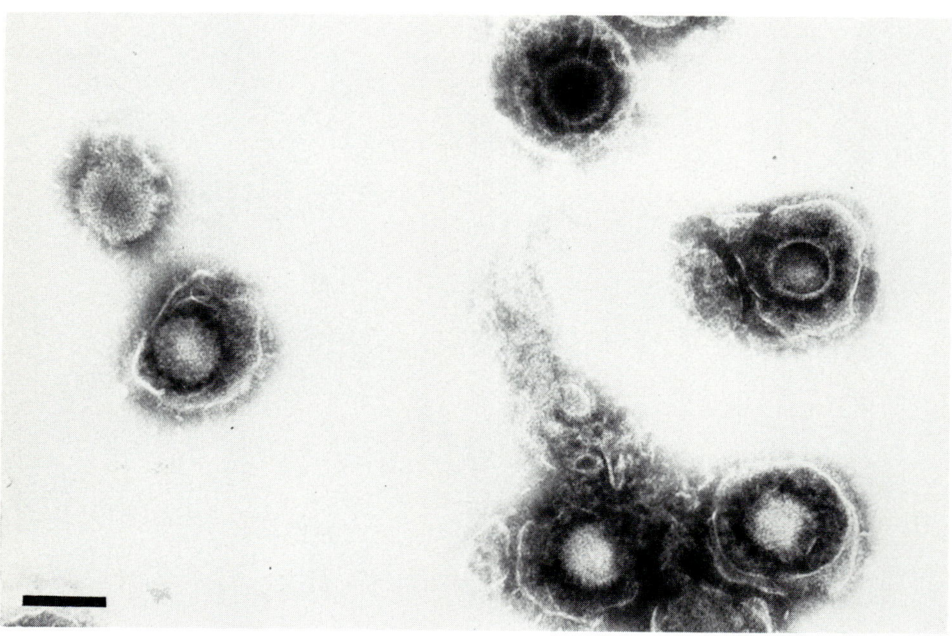

B

PLATE 6.   (A) Slightly distorted herpesvirus nucleocapsids among cell debris. Basic structure of particles is distorted but virus is still recognizable as herpesvirus species. (B) Varicella-zoster virus from vesicle
fluid obtained from a person with "chickenpox." All herpesviruses are morphologically identical.

Part VII
Complex DNA Viruses

Chapter 19

POXVIRIDAE

The poxviruses (pox = eruptive disease) are the largest and most complex of vertebrate viruses. There are a number of strains which are pathogenic for different animal species. These are divided into six genera. Two of these genera (*Orthopoxvirus* and *Parapoxvirus*) are important as pathogens of humans. Member viruses in each genus cross-react extensively in most conventional serological tests, but only one antigen is common to all genera. This antigen is considered to be a nucleoprotein complex. Variola virus, the causal agent of smallpox, is in the *Orthopoxvirus* genus. The saga of this disease is well known because its epidemic devastation helped to shape the history of the world. Once a destructor of entire populations, the disease is now believed to have been globally eradicated. Other poxviruses, such as camelpox and ectromelia, cause eruptive skin lesion diseases in lower animals. Only variola and molluscum contagiosum viruses are specific human pathogens. Other diseases such as Orf and cowpox are actually diseases of lower animals that may be transmitted to man. Except for viruses of the *Parapoxvirus* genus, all poxviruses are brick-shaped with a whorled-surface filament pattern, and are about 225 × 300 nm in size. The parapoxviruses (Orf and others) appear more ovoid with a regular spiraled filamentous surface pattern, and average only about 150 × 200 nm in size. The internal structure of all poxviruses studied has been found to be essentially identical.

Two types of poxvirus particles, originally called elementary bodies, are seen in tissue culture or vesicular fluids: brick-shaped particles, termed C particles, with a wide, clearly defined envelope and an electron-dense center, and mulberry (M) particles which have a whorled-surface filament pattern. The M particles usually have a central depression which represents the structure of the core or nucleoid surrounding a double-stranded DNA genome and the lateral bodies, which are attached at each side of the nucleoid. The function of these structures is not known, but when particles are intact the envelope pushes the bodies against the core so that the latter is dumbbell shaped. The lateral bodies and envelope can be removed by treating virions with detergent and 2-mercaptoethanol. The core then becomes expanded and brick-shaped.

REFERENCES

1. **Abdussalam, M.,** Elementary bodies of sheeppox, *Am. J. Vet. Res.,* 18, 614, 1957.
2. **Dales, S. and Siminovitch, L.,** The development of vaccinia virus in Earle's L strain cells as examined by electron microscopy, *J. Biophys. Biochem. Cytol.,* 10, 475, 1961.
3. **Dales, S. J.,** The uptake and development of vaccinia virus in L strain cells followed with labeled viral deoxyribonucleic acid, *J. Cell Biol.,* 18, 51, 1963.
4. **Dales, S. J. and Mosbach, E. H.,** Vaccinia as a model for membrane biogenesis, *Virology,* 35, 564, 1968.
5. **Easterbrook, K. B.,** Controlled degradation of vaccinia virions *in vitro:* an electron microscopic study, *J. Ultrastruct. Res.,* 37, 132, 1966.
6. **Epstein, M. A.,** Structural differentiation in the nucleoid of mature vaccinia virus, *Nature (London),* 181, 784, 1958.
7. **Hyde, J. M. and Peters, D.,** The organization of nucleoprotein within fowlpox virus, *J. Ultrastruct. Res.,* 35, 626, 1971.
8. **Joklik, W. K.,** The poxviruses, *Bacteriol. Rev.,* 30, 33, 1966.
9. **Medzon, E. L. and Bauer, H.,** Structural features of vaccinia virus revealed by negative staining, *Virology,* 40, 860, 1970.
10. **Mitchiner, M. B.,** The envelope of vaccinia and orf viruses: an electron-cytochemical investigation, *J. Gen. Virol.,* 5, 211, 1969.

11. **Moss, B.,** Reproduction of poxviruses, in *Comprehensive Virology,* Vol. 3, Fraenkel-Conrat, H. and Wagner, R. R., Eds., Plenum Press, New York, 1974, 405.

12. **Nagington, J. and Horne, R. W.,** Morphological studies of orf and vaccinia viruses, *Virology,* 16, 248, 1962.

13. **Nagington, J., Newton, A. A., and Horne, R. W.,** The structure of orf virus, *Virology,* 23, 461, 1964.

14. **Noyes, W. F.,** The surface fine structure of vaccinia virus, *Virology,* 17, 282, 1962.

15. **Noyes, W. F.,** Further studies on the structure of vaccinia virus, *Virology,* 18, 511, 1962.

16. **Peters, D.,** Morphology of resting vaccinia virus, *Nature (London),* 178, 1453, 1956.

17. **Peters, D. and Muller, G.,** The fine structure of the DNA containing core of vaccinia virus, *Virology,* 21, 266, 1963.

18. **Pogo, B. G. T. and Dales, S.,** Two deoxyribonuclease activities within purified vaccinia virus, *Proc. Natl. Acad. Sci. USA,* 63, 820, 1969.

19. **Westwood, J. C. N., Harris, W. J., Zwartouw, H. T., Titmuss, D. H. J., and Appleyard, G.,** Studies on the structure of vaccinia virus, *J. Gen. Microbiol.,* 34, 67, 1964.

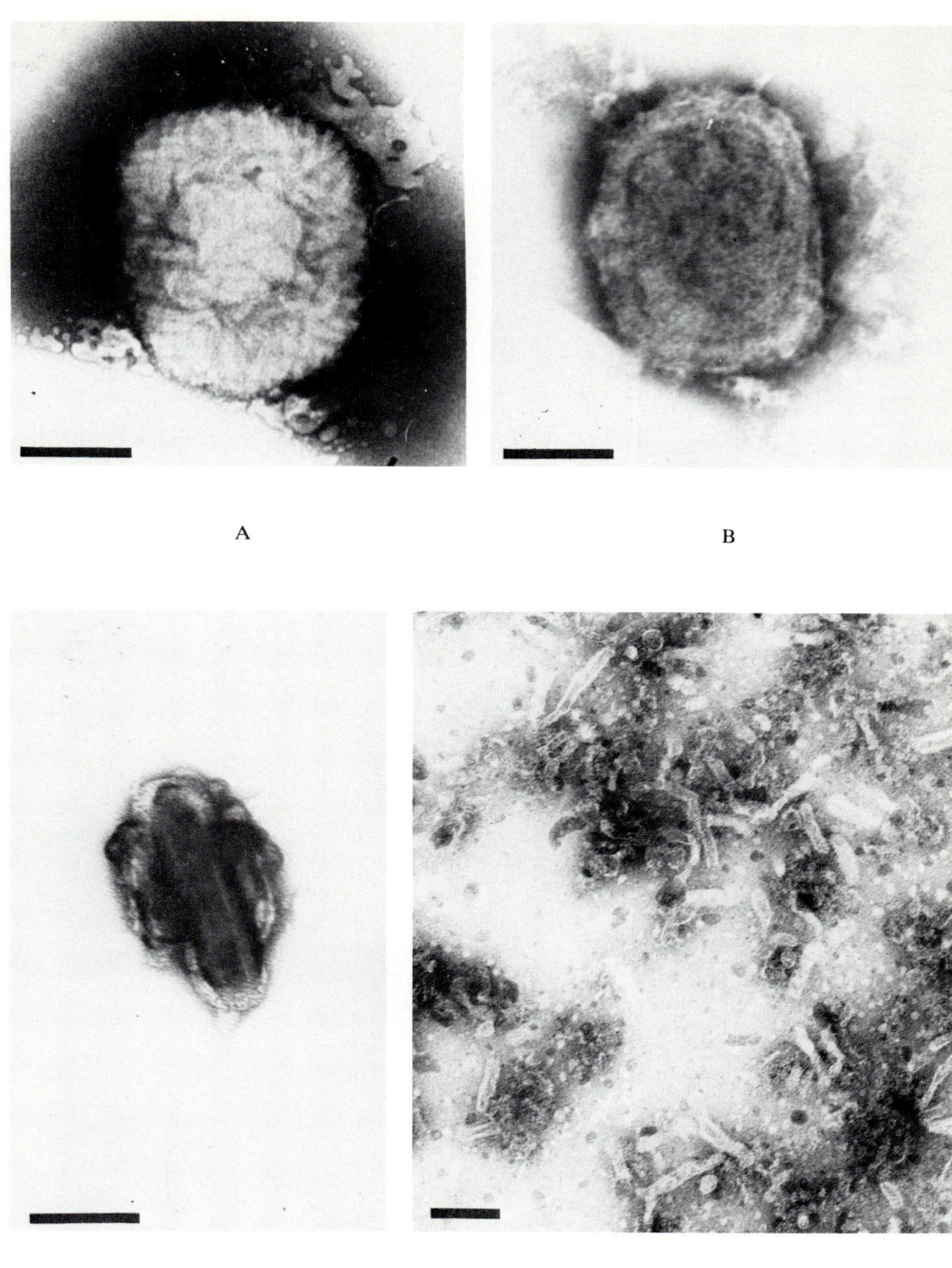

A

B

C

D

PLATE 1. (A) ''M'' form of poxvirus. (B) ''C'' form of poxvirus. (C) Poxvirus core with attached lateral bodies. (D) Surface tubules isolated from vaccinia virus. All bars equal 100 nm.

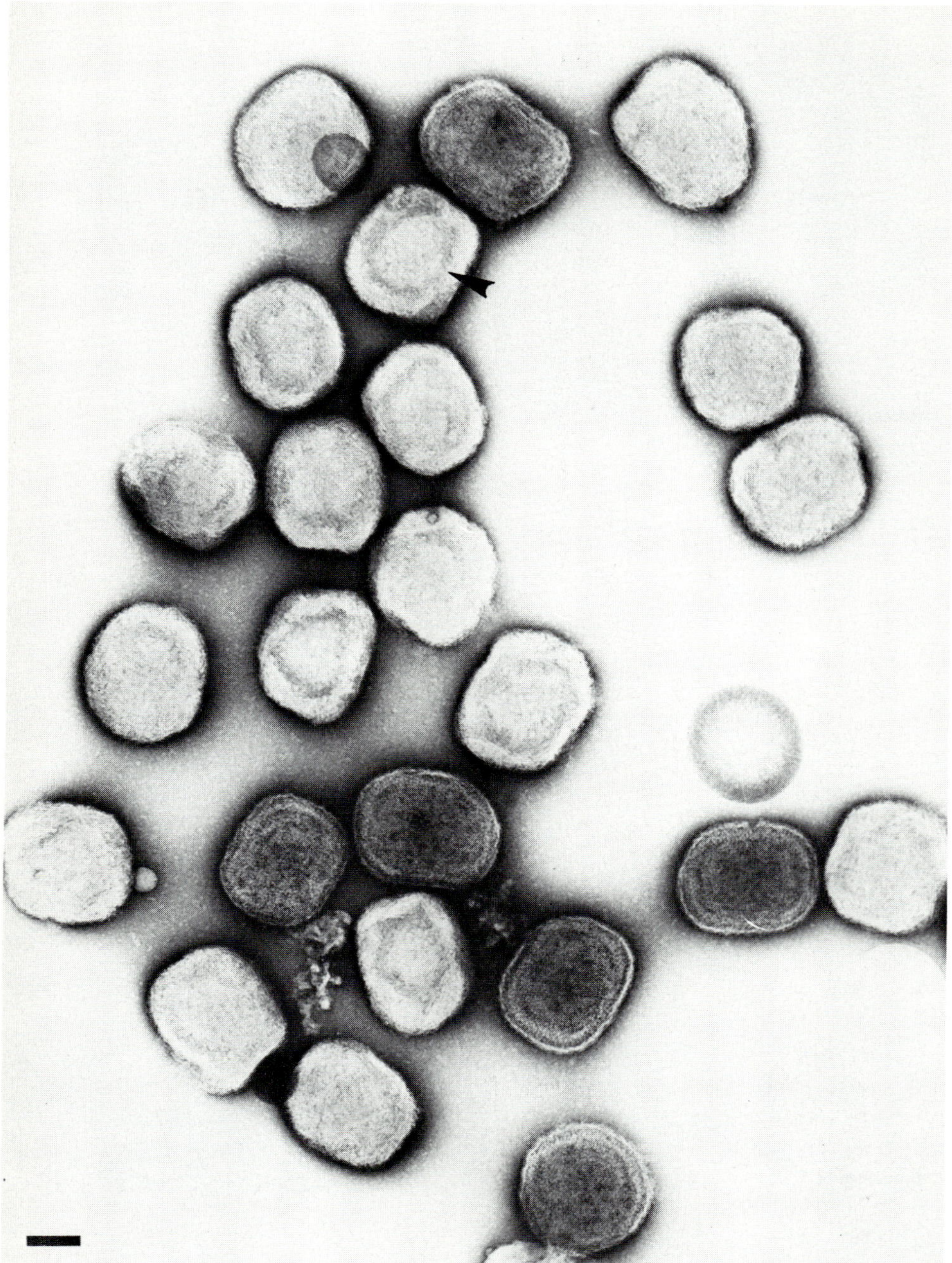

PLATE 2.   Group of vaccinia virus particles. Some are penetrated with stain and the envelope is visible. Depressions in the center of some particles (arrow) result from lateral bodies which are on two sides of the virus core.

PLATE 3.   (A) "M" and "C" forms of Orf virus from vesicular fluid. (B) Orf virus from a skin lesion. (C) Higher magnification of Orf virus . Orf virus is more ovoid than other poxviruses and can easily be identified by EM.

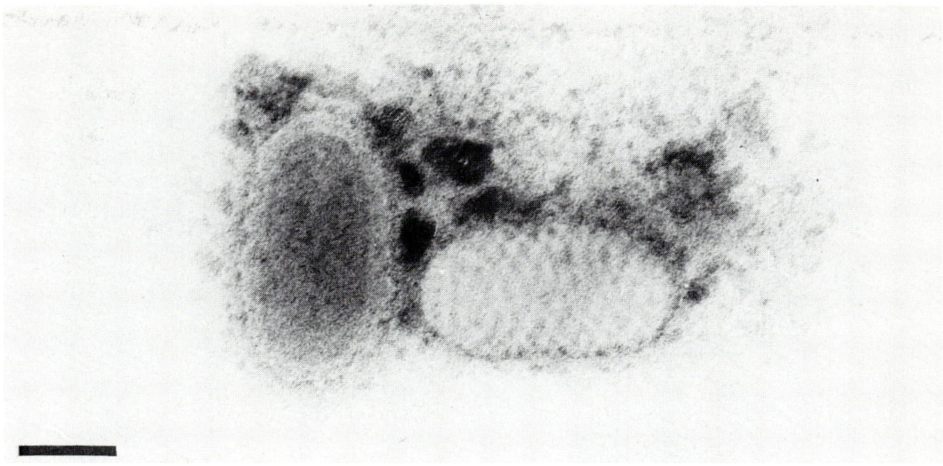

A

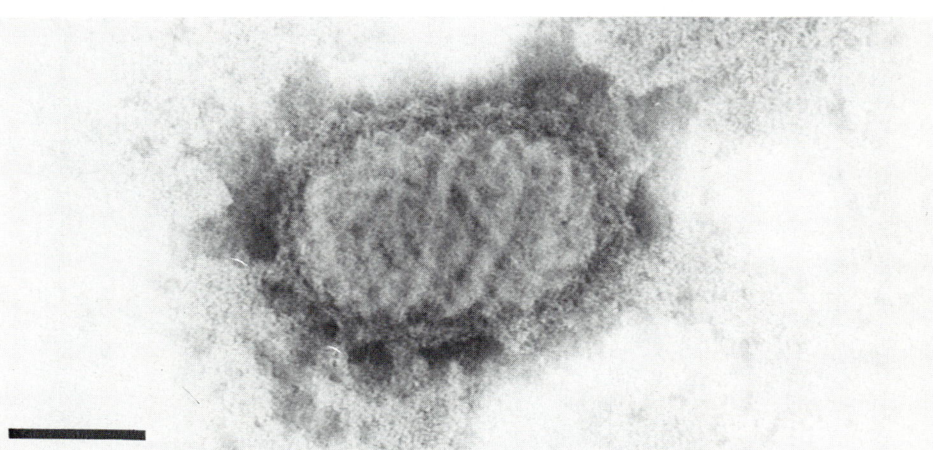

B

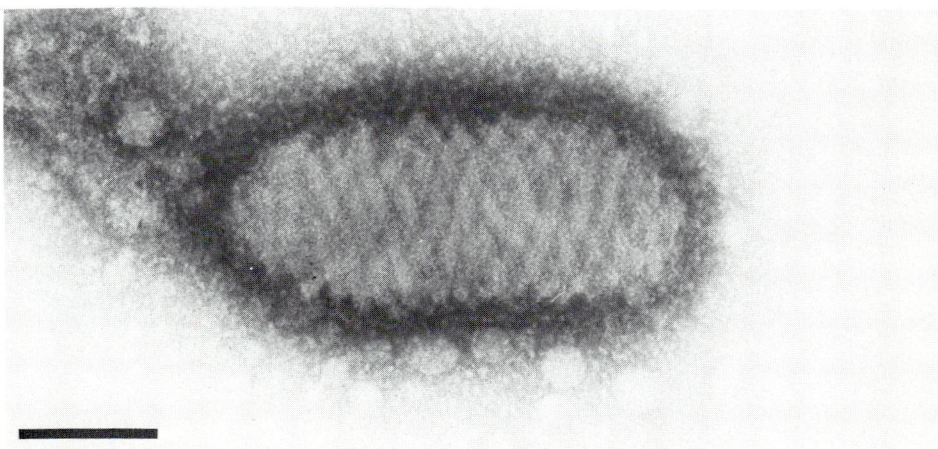

C

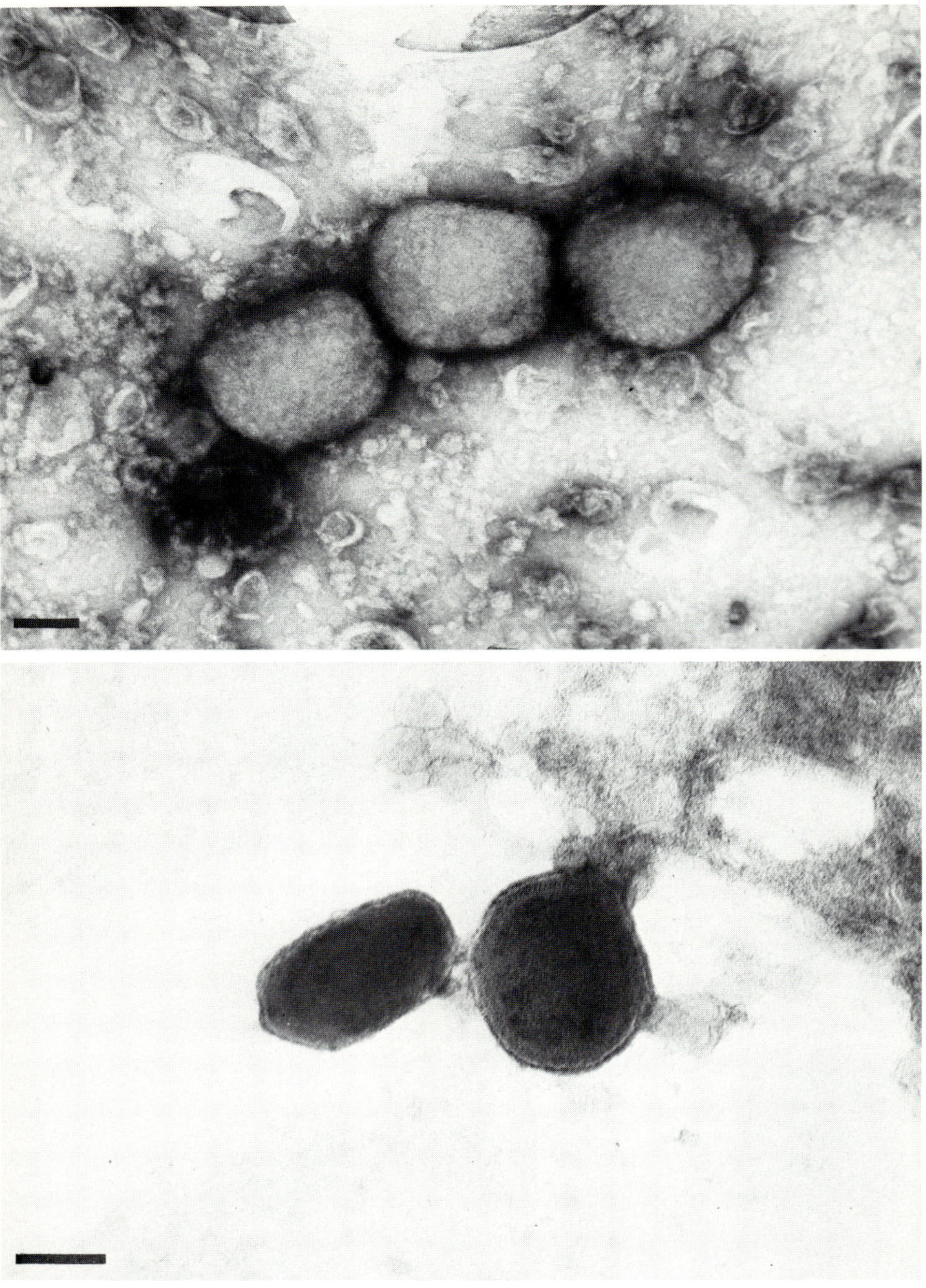

PLATE 4.   Smallpox virus from human smallpox vesicular fluid.

Part VIII
Viruses with Unknown Symmetry

Chapter 20

HEPATITIS A AND B VIRUSES

Infectious (short-incubation) hepatitis is caused by hepatitis A virus (HAV). The virus primarily affects the young and is readily transmitted by oral and parenteral routes. Virus has been detected in stools of acutely ill persons, and a causal relationship between the virus and infectious hepatitis has been established by IEM. HAV particles are 27 nm in diameter and appear angular. Little is known about the structure of HAV because visualization of the virus has been limited. Whether the virus has an RNA or DNA genome also remains to be elucidated. Taxonomically, HAV is currently considered as a possible member of the Parvoviridae. The virus is quite different from hepatitis B virus (HBV), which is a DNA-containing virus that causes serum (long-incubation) hepatitis, and there is no cross-immunity between these diseases.

HBV (Dane particle) has a mean diameter of 42 nm and has a double-stranded DNA genome which is surrounded by a core. The virus may be transmitted by the parenteral route by infected blood or by the fecal-oral route. Virus can also be found in feces and saliva. Presence of hepatitis-associated antigen (HAA or Australian antigen) is generally a fairly good marker for presence of HBV. The HAA is a surface antigen of the infectious particle which may be present in large amounts in serum. HAA is non-infectious and contains no nucleic acid. Presence of the antigen can be detected by EM and a variety of serological tests. HAA is about 20 nm in diameter. Filaments of antigen can also sometimes be seen in antigen aggregates.

REFERENCES

1. **Almeida, J. D., Waterson, A. P., Trowell, J. M. and Neal, G.,** The finding of virus-like particles in two Australia-antigen-positive human livers, *Microbios,* 2, No. 6, 145, 1970.
2. **Almeida, J. D., Rubenstein, D., and Stott, J. E.,** New antigen-antibody system in Australia-antigen-positive hepatitis, *Lancet,* 2, 1225, 1971.
3. **Bayer, M. E., Blumberg, B. S., and Werner, B.,** Particles associated with Australia antigen in the sera of patients with leukaemia, Down's syndrome and hepatitis, *Nature (London),* 218, 1057, 1968.
4. **Dane, D. S., Cameron, C. H., and Briggs, M.,** Virus-like particles in serum of patients with Australia-antigen-associated hepatitis, *Lancet,* 1, 695, 1970.
5. **Feinstone, S. M., Kapikian, A. Z., and Purcell, R. H.,** Hepatitis A: detection by immune electron microscopy of a virus-like antigen associated with acute illness, *Science,* 182, 1026, 1973.
6. **Feinstone, S. M., Moritsugu, Y., Shih, J. W. K., Gerin, J. L., and Purcell, R. H.,** Characterization of hepatitis A virus, in *Viral Hepatitis: A Contemporary Assessment of Etiology, Epidemiology, Pathogenesis and Prevention,* Vyas, G. N., Cohen, S. N., and Schmid, R., Eds., Franklin Institute Press, Philadelphia, 1978, 41.
7. **Gerin, G. L.,** Structure of hepatitis B antigen (HB$_s$Ag), in *Mechanisms of Virus Disease,* Robinson, W. S. and Fox, C. R., Eds., Benjamin, Menlo Park, Calif., 1974, 215.
8. **Gerin, J. L., Holland, P. V., and Purcell, R. H.,** Australia antigen: large scale purification from human serum and biochemical studies of its proteins, *J. Virol.,* 7, 569, 1971.
9. **Hoggan, M. D.,** Small DNA viruses, in *Comparative Virology,* Maramorosch, K. and Kurstak, E., Eds., Adacemic Press, New York, 1971, 43.
10. **Robinson, W. S.,** Viruses of human hepatitis A and B, in *Comprehensive Virology,* Vol. 14, Fraenkel-Conrat, H. and Wagner, R. R., Eds., Plenum Press, New York, 1979, 471.
11. **Robinson, W. S. and Lutwick, L. I.,** The virus of hepatitis, type B, *N. Engl. J. Med.,* 295, 1168, 1976.
12. **Siegl, G. and Frosner, G. G.,** Characterization and classification of virus particles associated with hepatitis A. I. Size, density and sedimentation, *J. Virol.,* 26, 40, 1978.

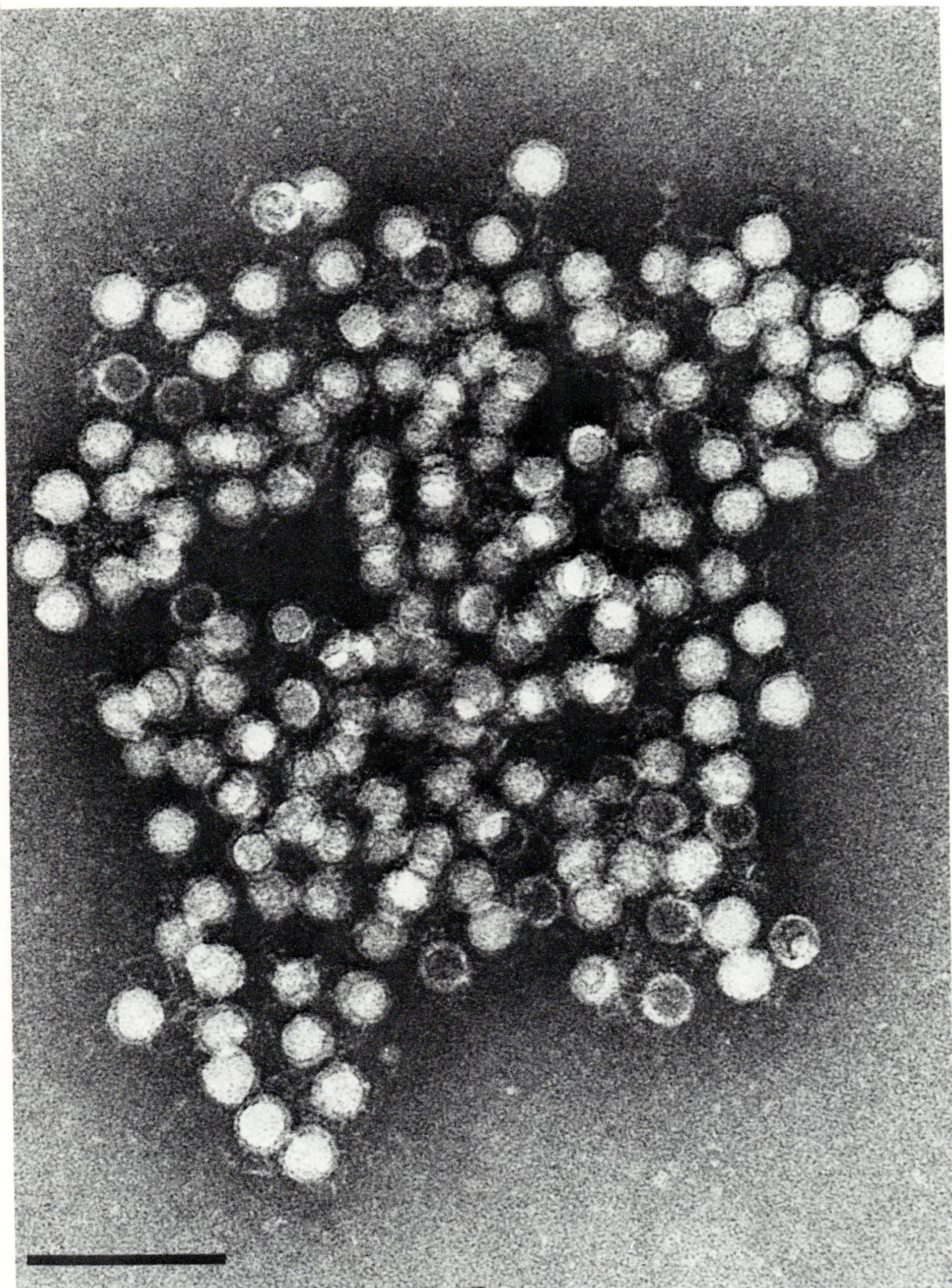

PLATE 1.   Hepatitis A virus from chimpanzee liver preparation treated with protease and banded in CsCl. All bars equal 100nm.

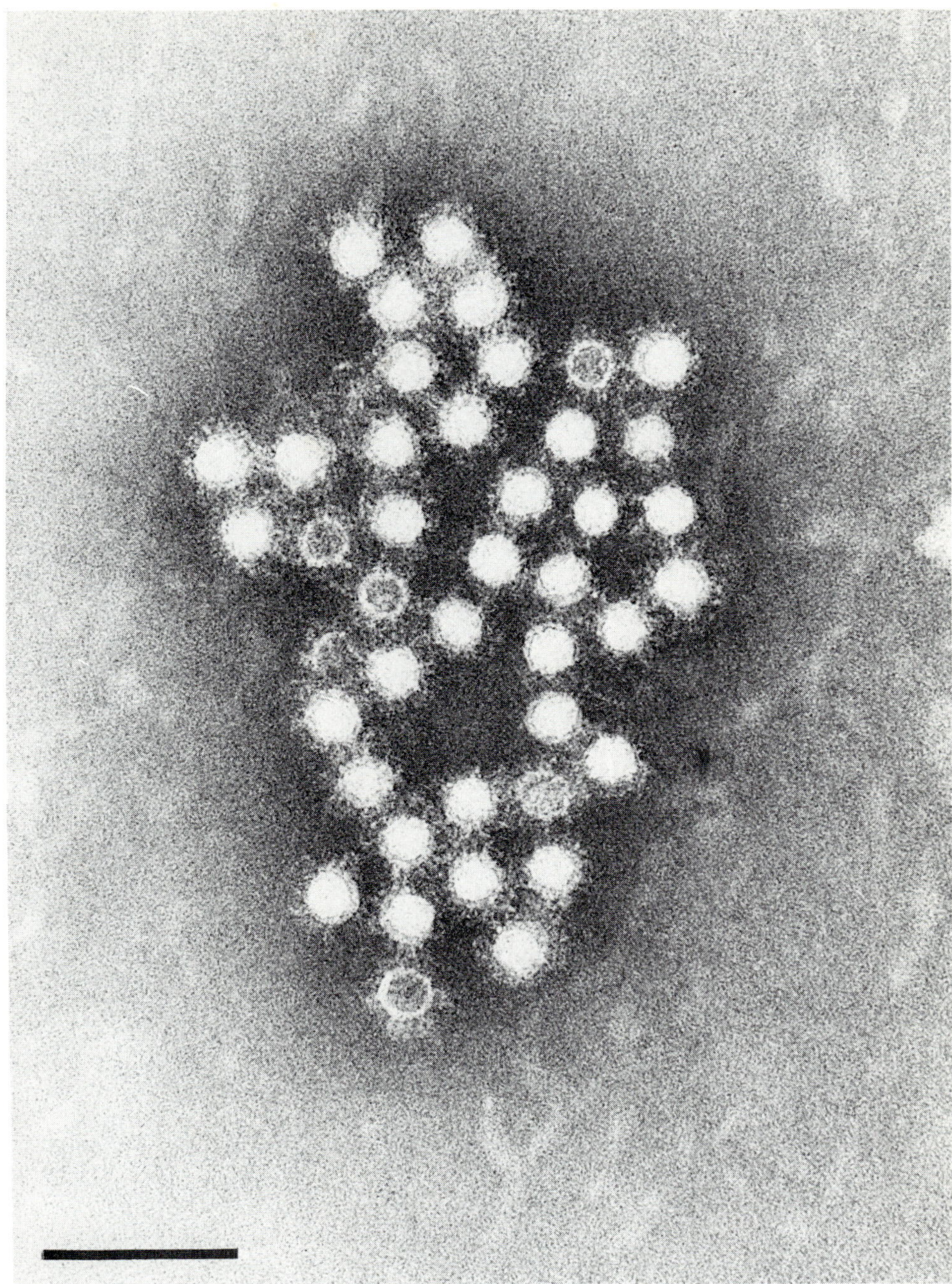

PLATE 2. IEM of Hepatitis A virus from marmoset liver preparation banded in CsCl. IEM is against human anti-HAV.

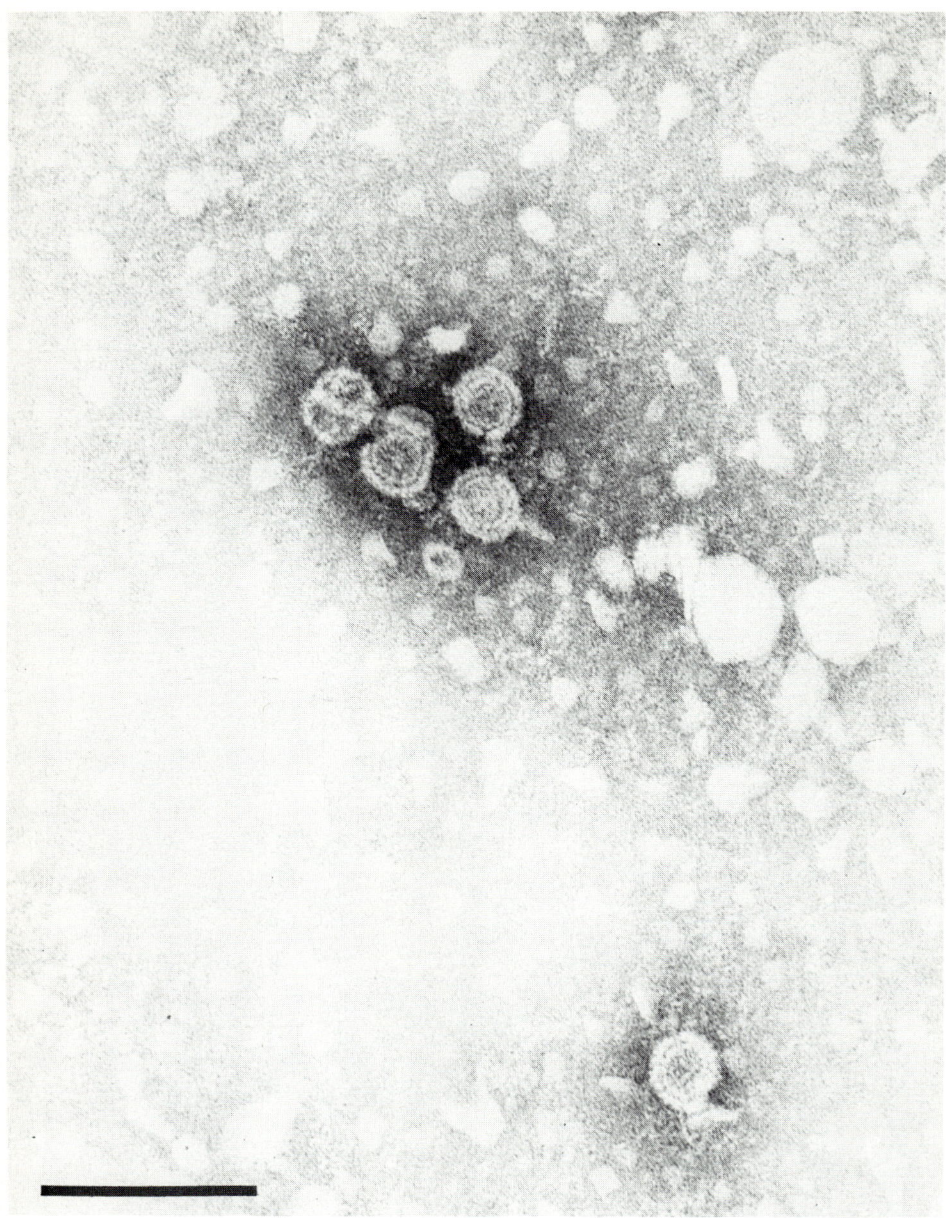

PLATE 3.   Hepatitis B Dane particles.

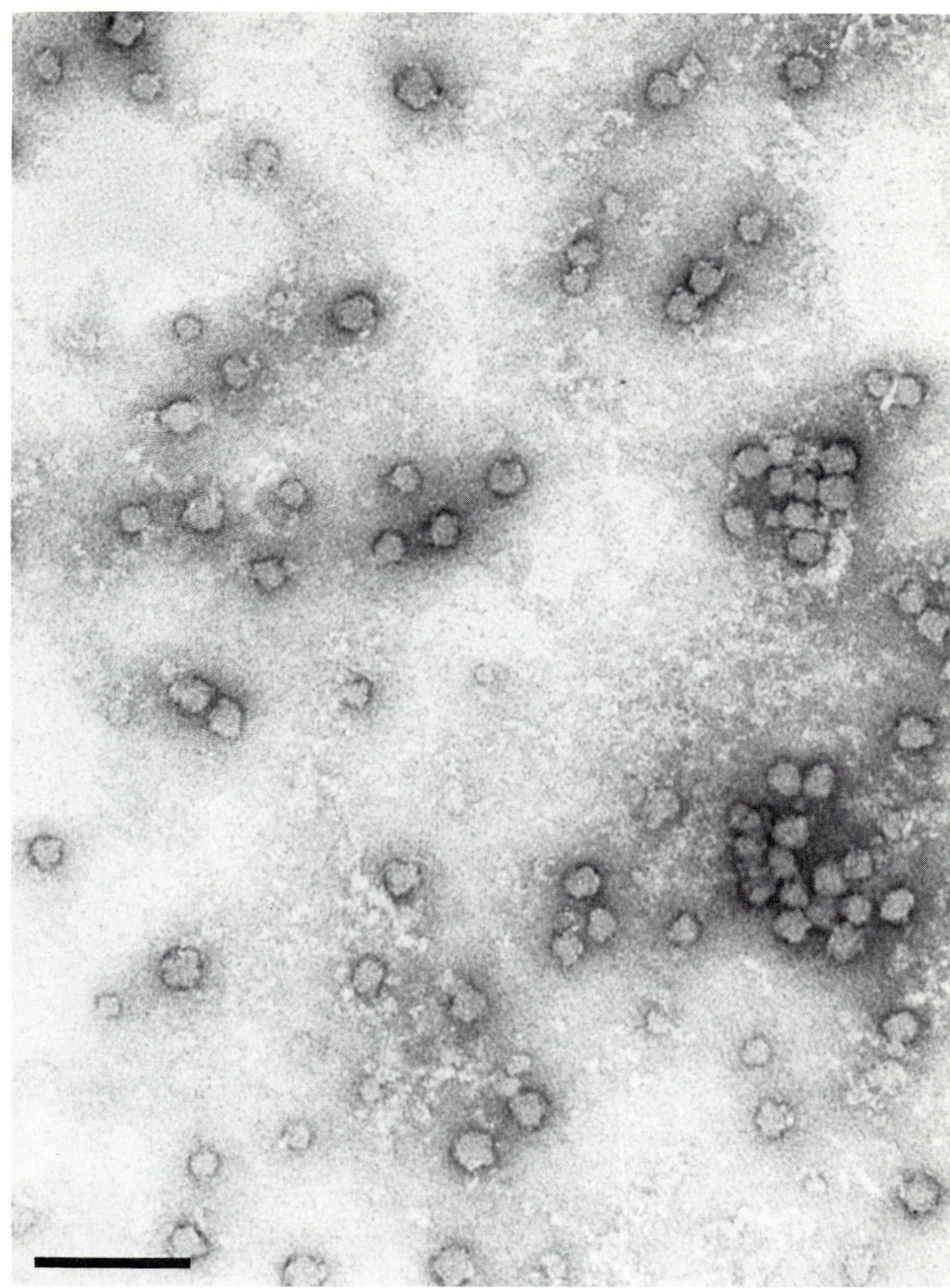

PLATE 4.   Hepatitis-associated antigen (Australia antigen) from serum.

*Part IX
Miscellaneous Particles*

Chapter 21

MISCELLANEOUS VIRUS-LIKE PARTICLES AND BACTERIOPHAGE

The most common problem in identifying the small icosahedral viruses in tissue culture preparations is discriminating these particles from ribosomes, especially when the viruses are not in groups and no penetrated particles are evident. Ribosomes are less angular and more variable in size than picornaviruses or parvoviruses. The small size of all these entities (20 to 30nm), however, makes it difficult to see definitive shape and discern one size from another. We, therefore, look for groups of three or four particles with the same size and shape before making a definitive diagnosis. No problems occur in identifying the larger icosahedral viruses. Each has a characteristic structure and the particles are not easily confused with cell debris.

Some of the enveloped helical viruses may be difficult to identify among cell debris because cell membrane fragments are often "virus-shaped" and have surface projections. The structures also pose a significant problem when trying to identify viruses in material from diseases of unknown etiology. Collagen is also present in some tissue, egg material, stools, and in tissue cultures inoculated with muscle tissue. It has a definite striated appearance and is easily recognized among cell debris.

Numerous viruses and virus-like structures are seen in stool preparations. It is not difficult to identify rotaviruses, adenoviruses, calicivirus, and coronavirus in stools. However, identifying small viruses associated with diarrhea, hepatitis A virus, and agents from diseases of unknown etiology must be done by IEM because so many small phage and virus-like particles are present in stools. Most phages are easily recognized as such, but those that are very small may be confused with enteroviruses or parvoviruses.

REFERENCES

1. **Bradley, D. E.,** A comparative study of the structural and biological properties of bacteriophages, in *Comparative Virology,* Maramorosch, K. and Kurstak, E., Eds., Academic Press, New York, 1971, 207.
2. **Brenner, S., Streisinger, G., Horne, R. W., Champe, S. P., Barnett, L., Benzer, S., and Reese, M. W.,** Structural components of bacteriophage, *J. Mol. Biol.,* 1, 281, 1959.
3. **Tikhonenko, A. S.,** *Ultrastructure of Bacterial Viruses,* Plenum Press, New York, 1970.

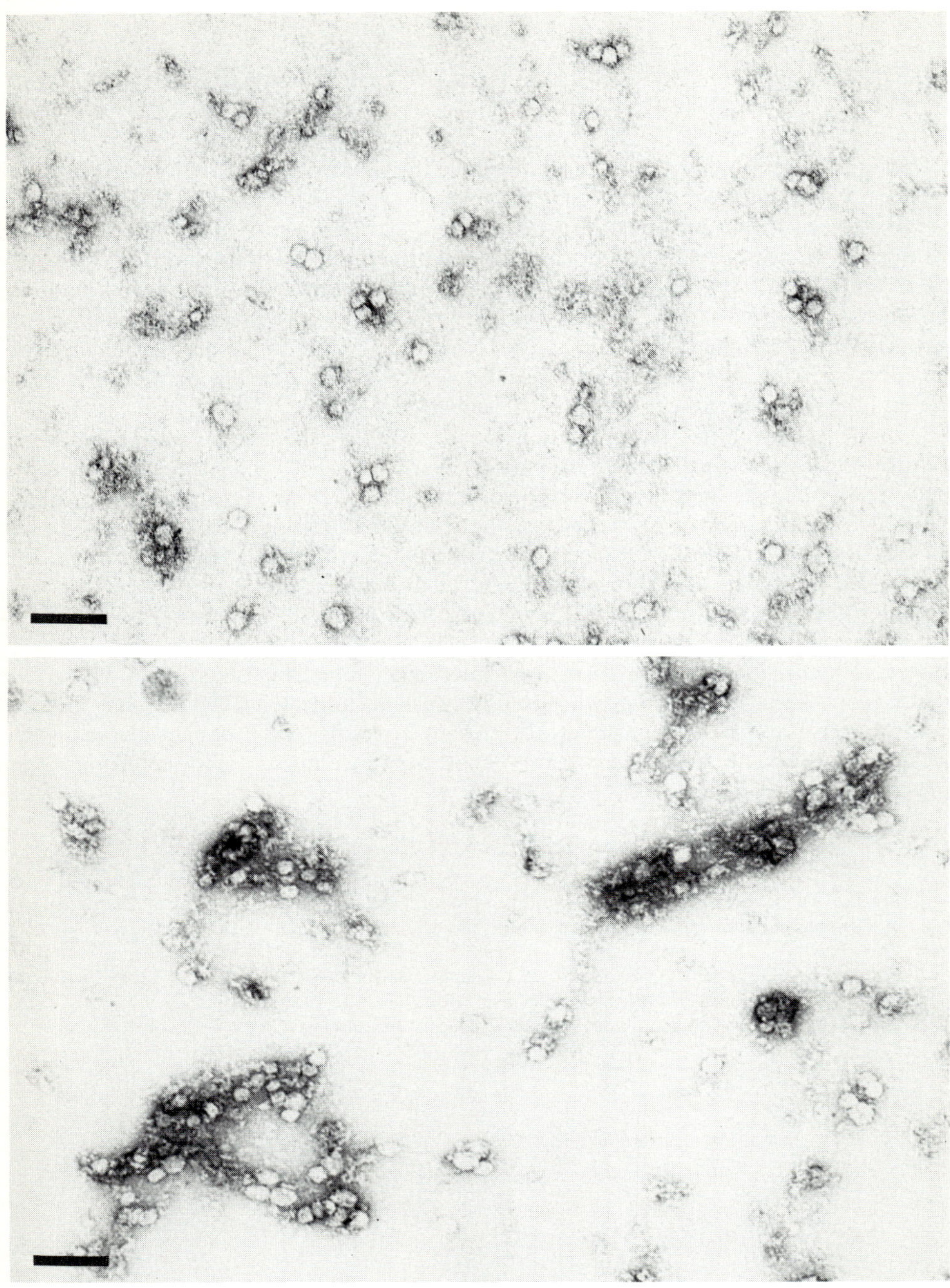

PLATE 1.   Ribosomes and polyribosomes from VERO cells. All bars equal 100 nm.

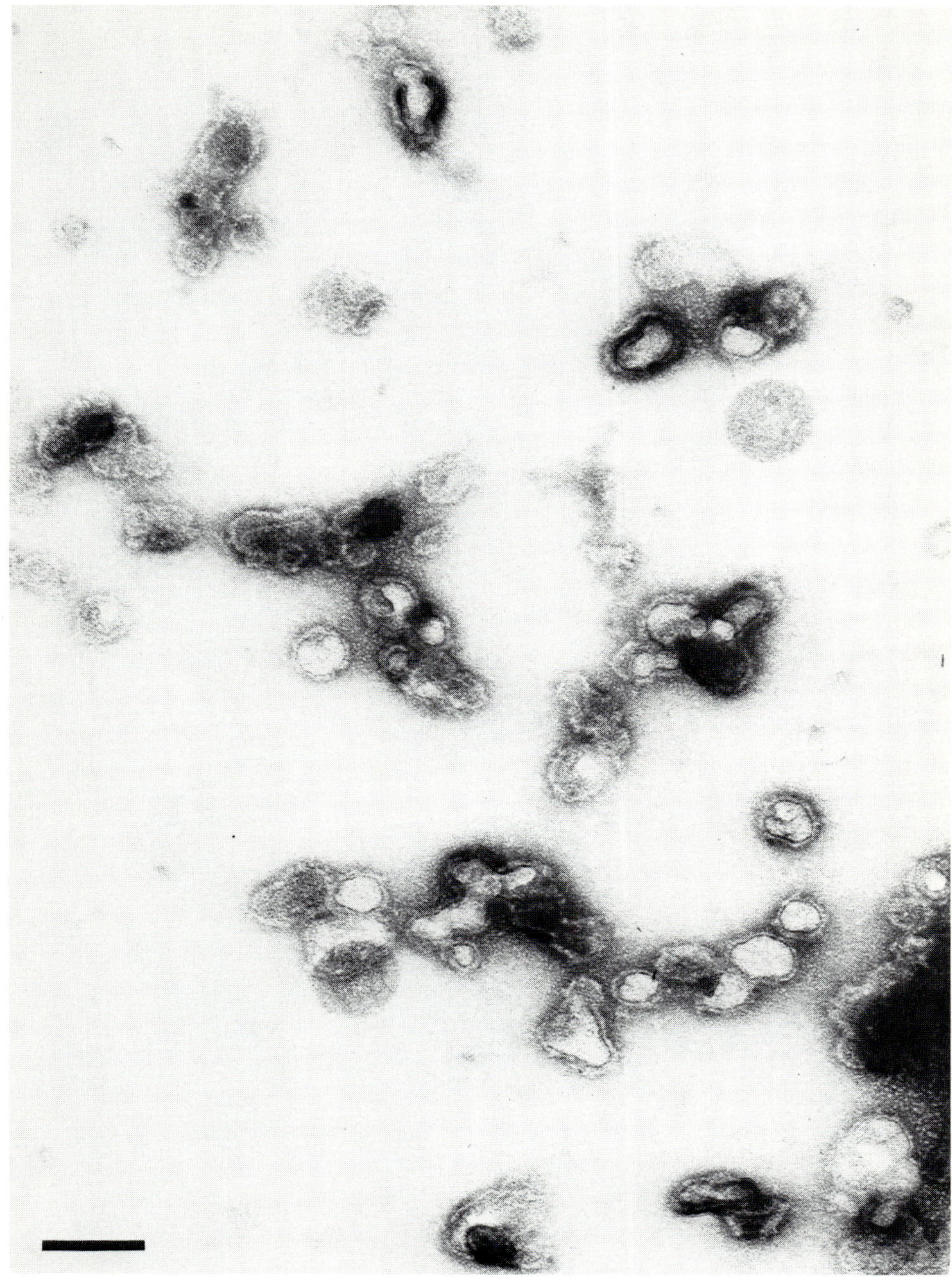

PLATE 2. Tissue culture cell membrane fragments. These may be round or pleomorphic and frequently
have well-defined surface projections resembling those of orthomyxoviruses.

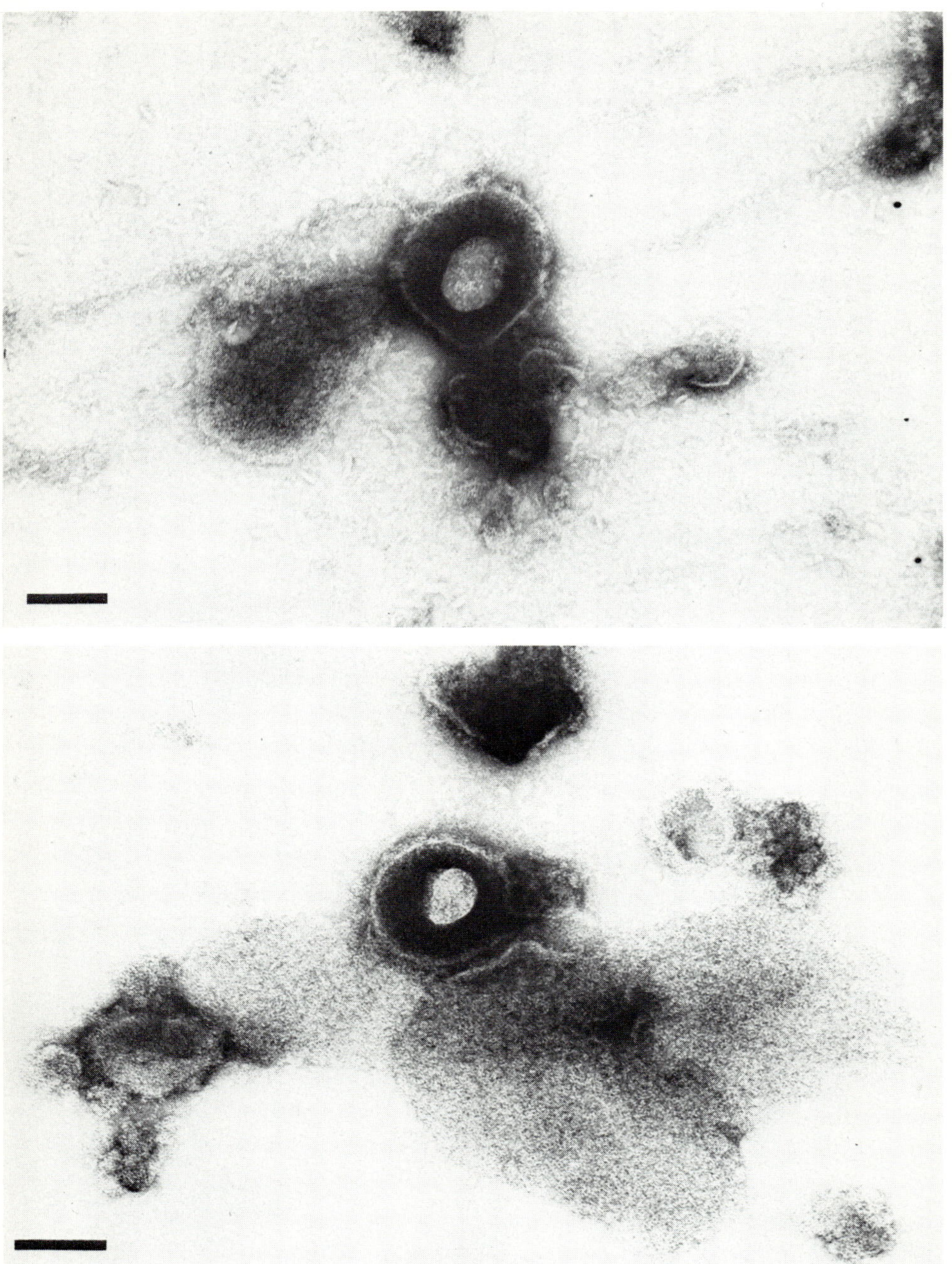

PLATE 3.   Artifactual virus-like particles from tissue culture which resemble herpesviruses.

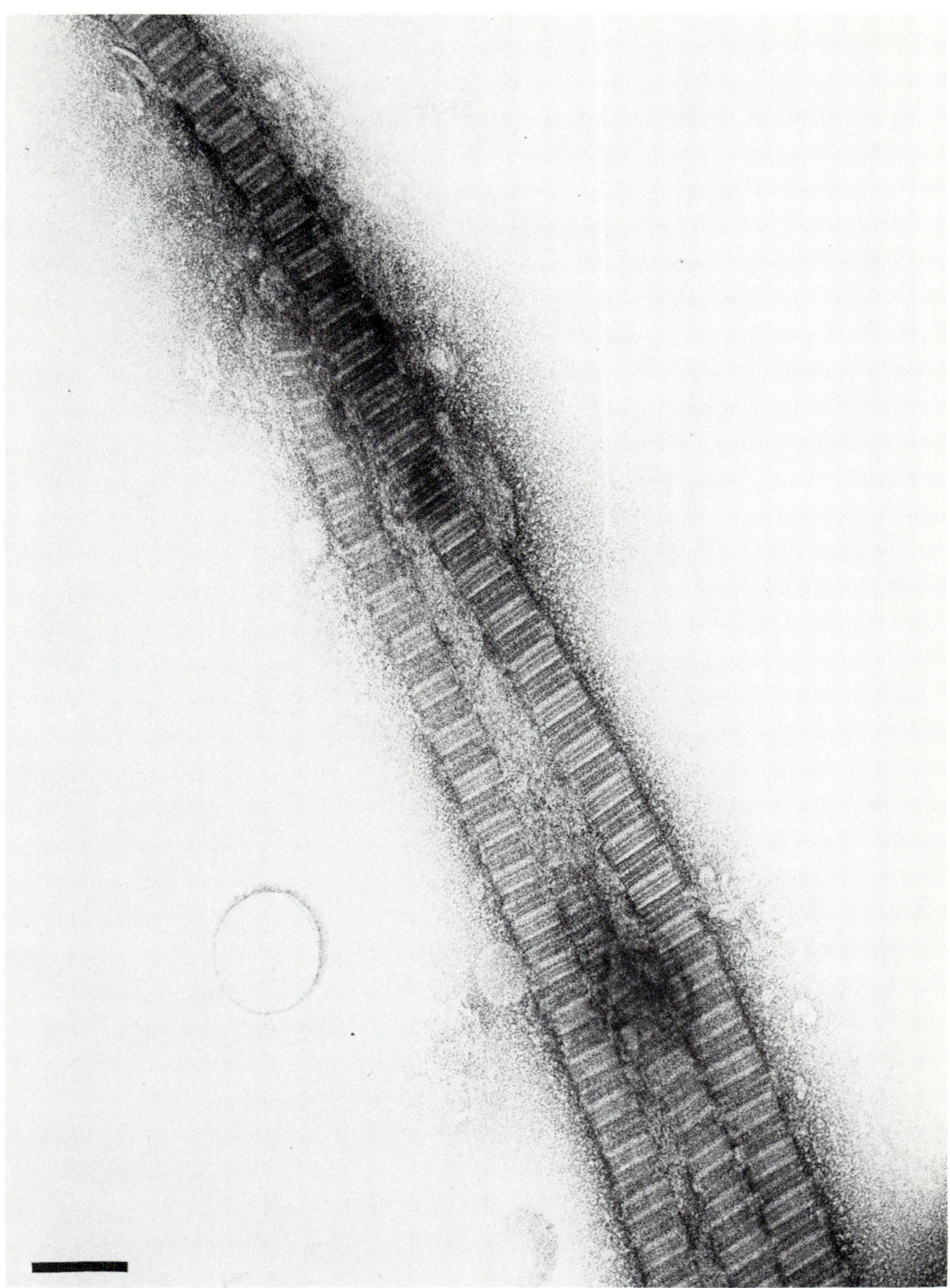

PLATE 4.   Collagen from first passage of lung biopsy in tissue culture.

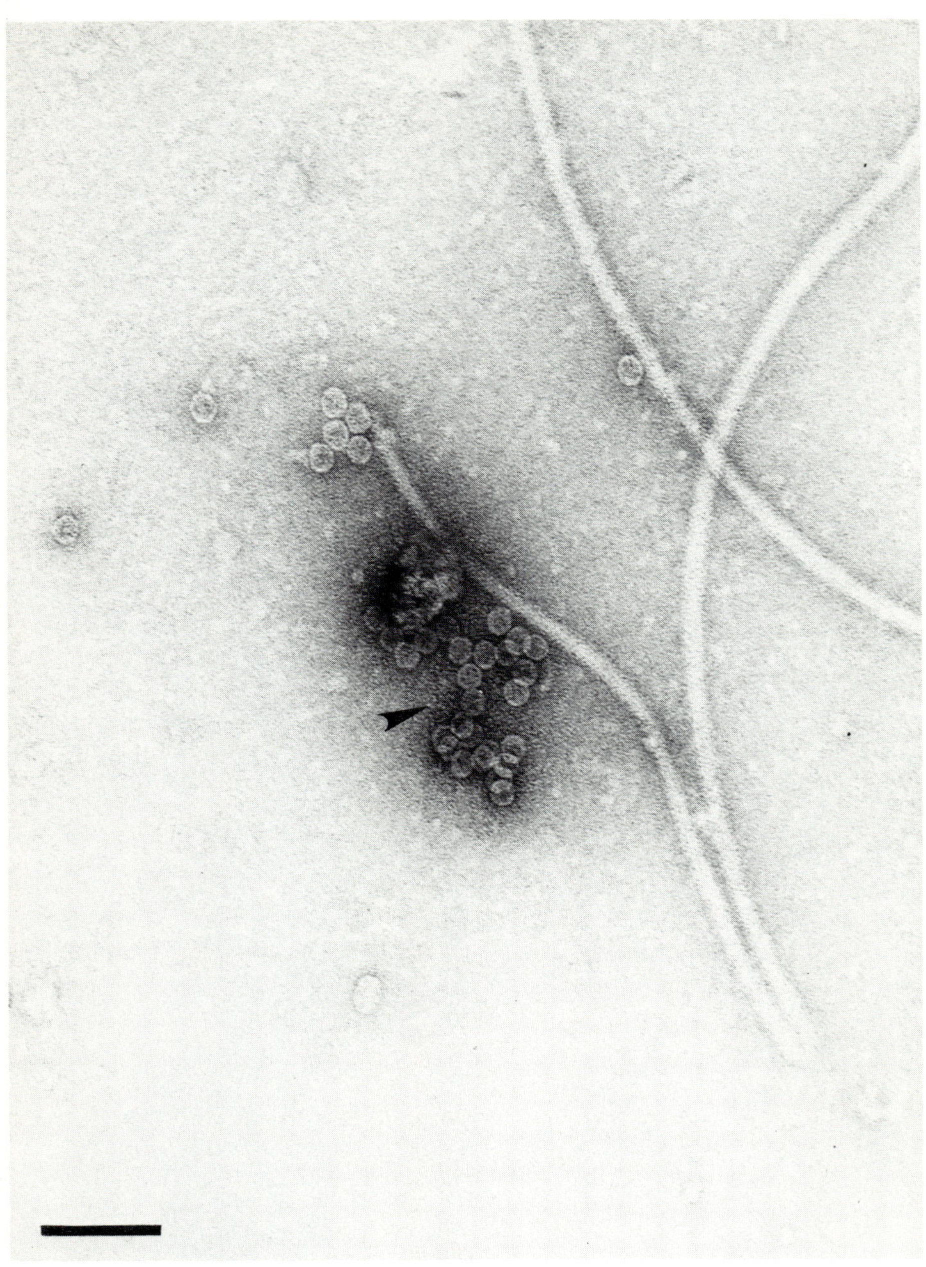

PLATE 5.   Bacteriophage MS (arrow) and flagella.

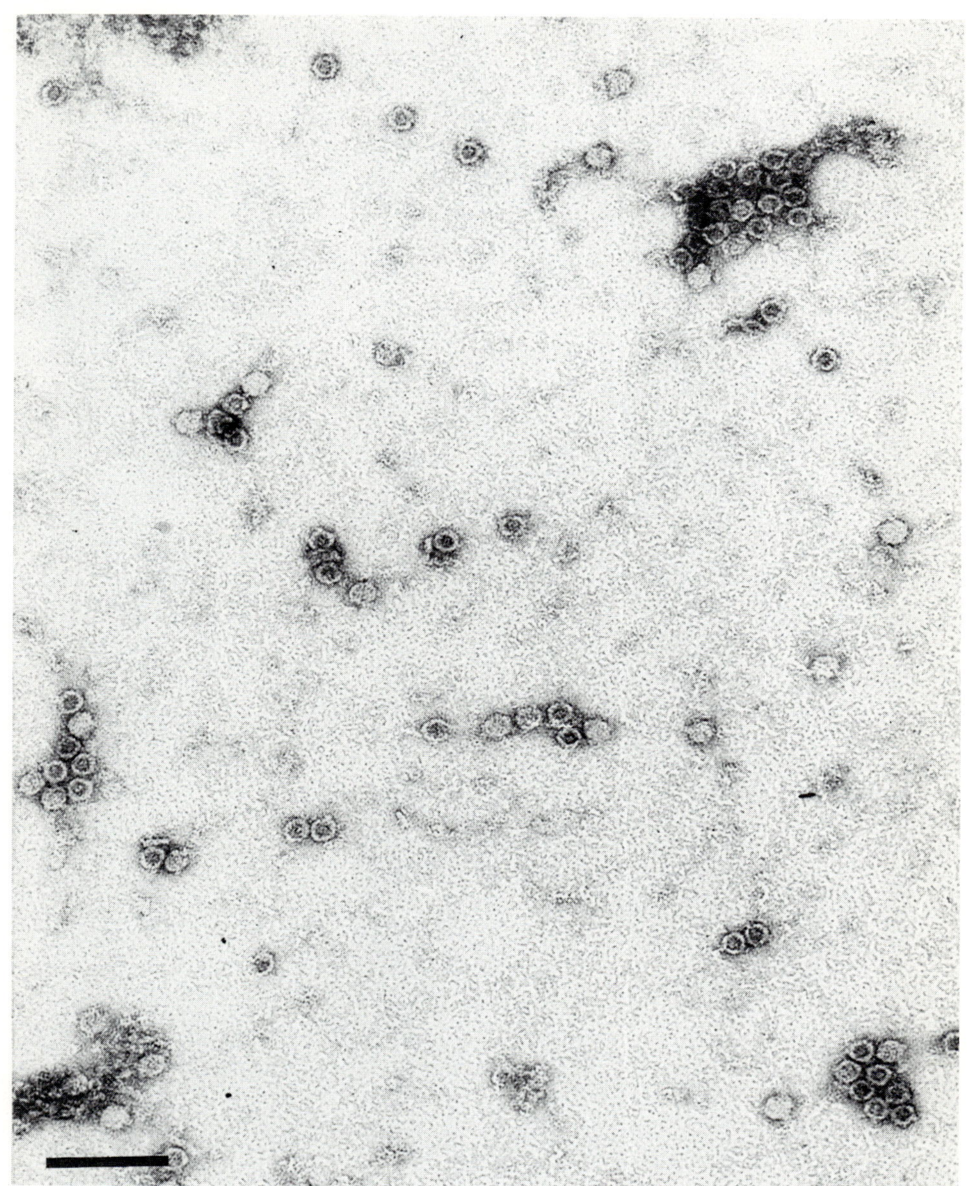

PLATE 6.   Bacteriophage ϕ × 174.

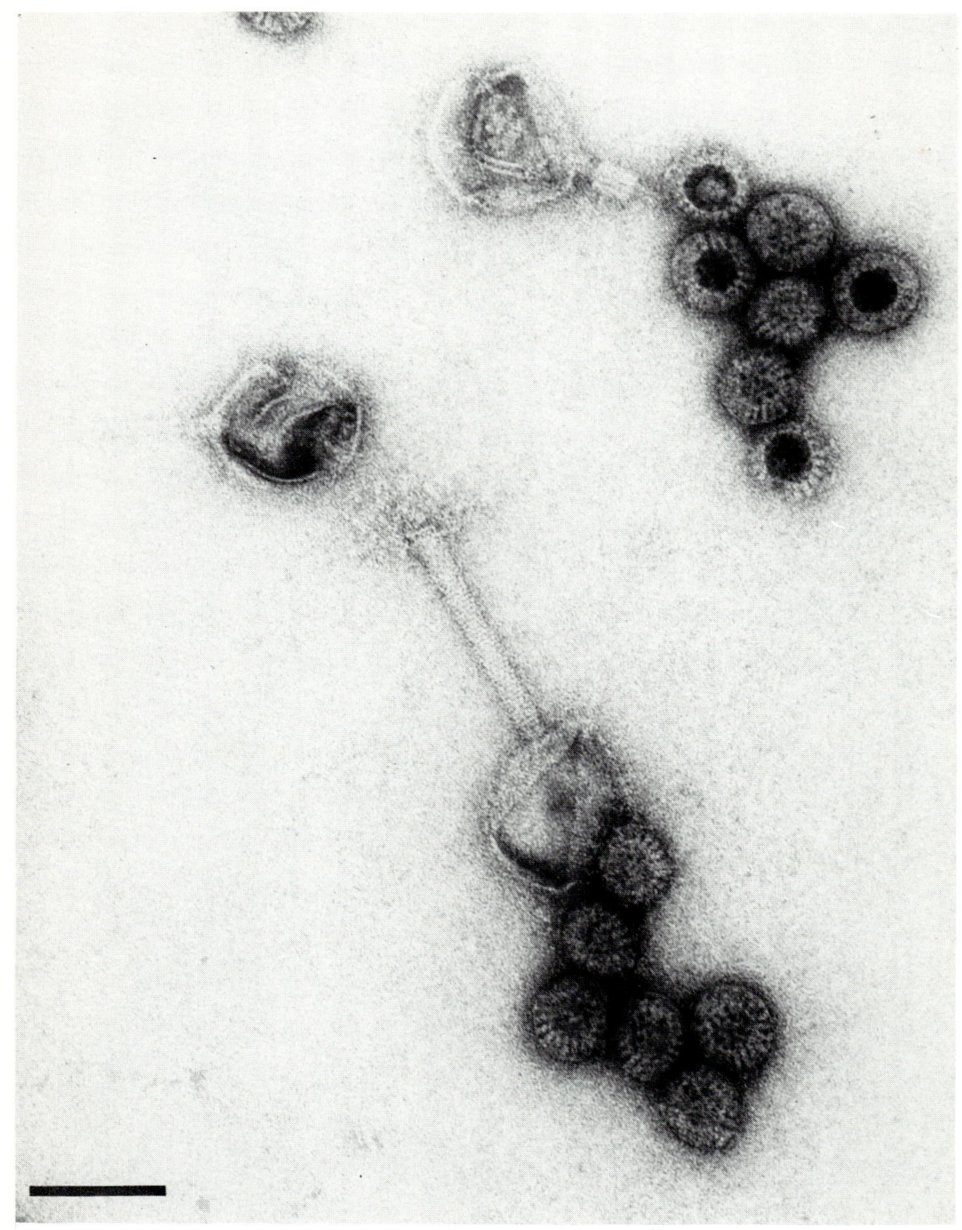

PLATE 7.   T-even phage and rotavirus.

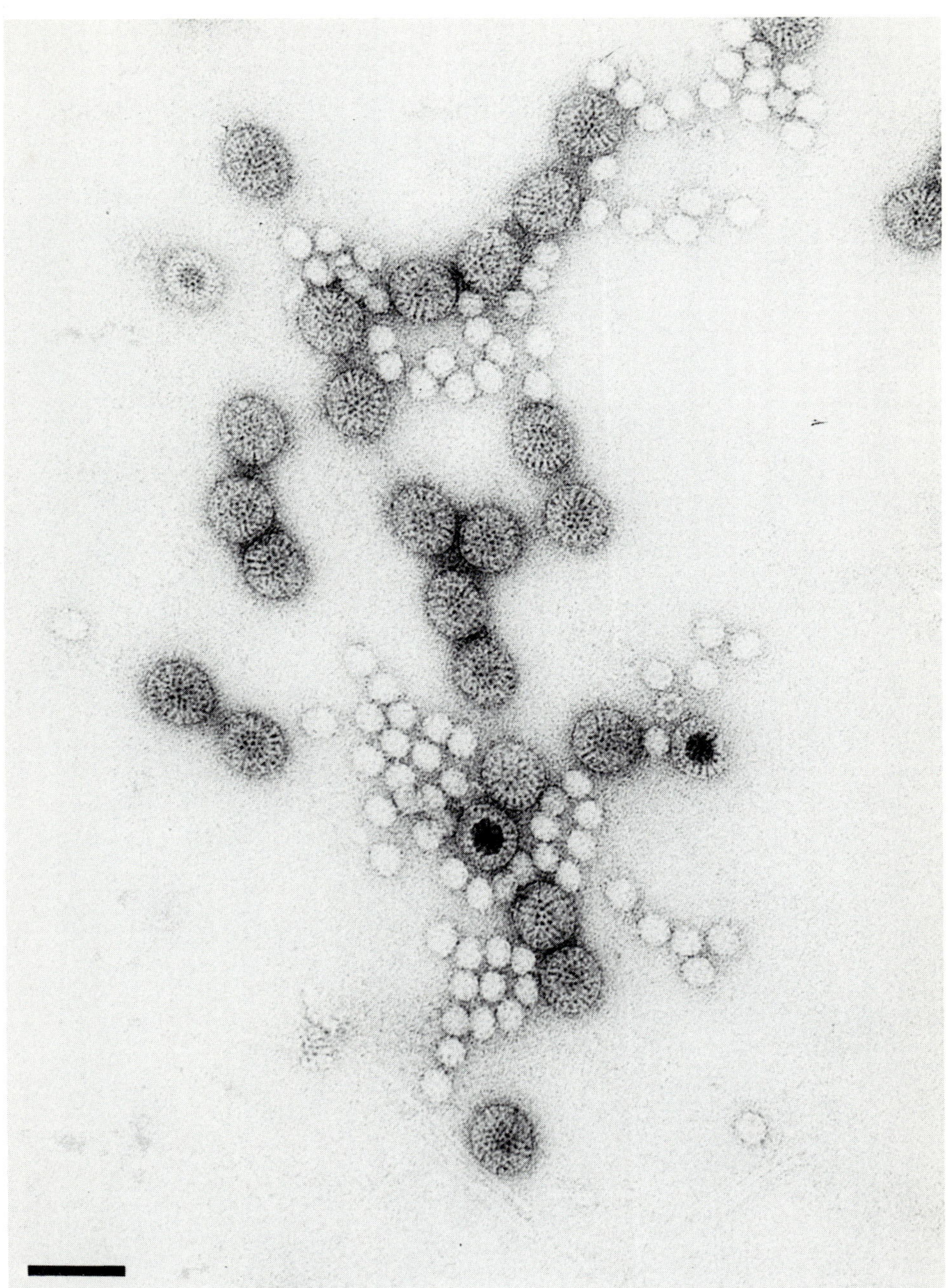

PLATE 8. Rotavirus and unidentified small round virus from stool.

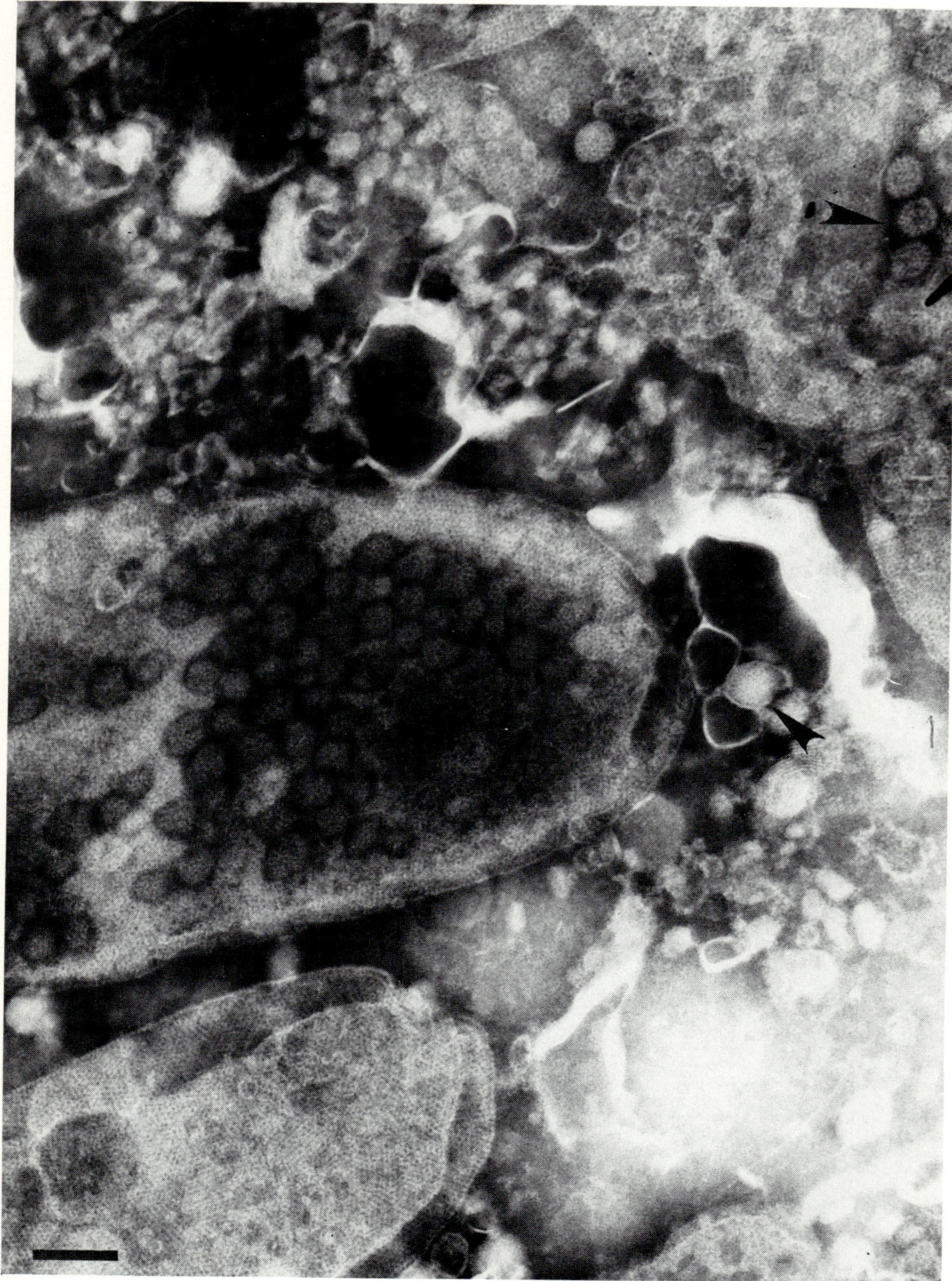

PLATE 9.   Bacteria with surface phage and rotaviruses (arrow).

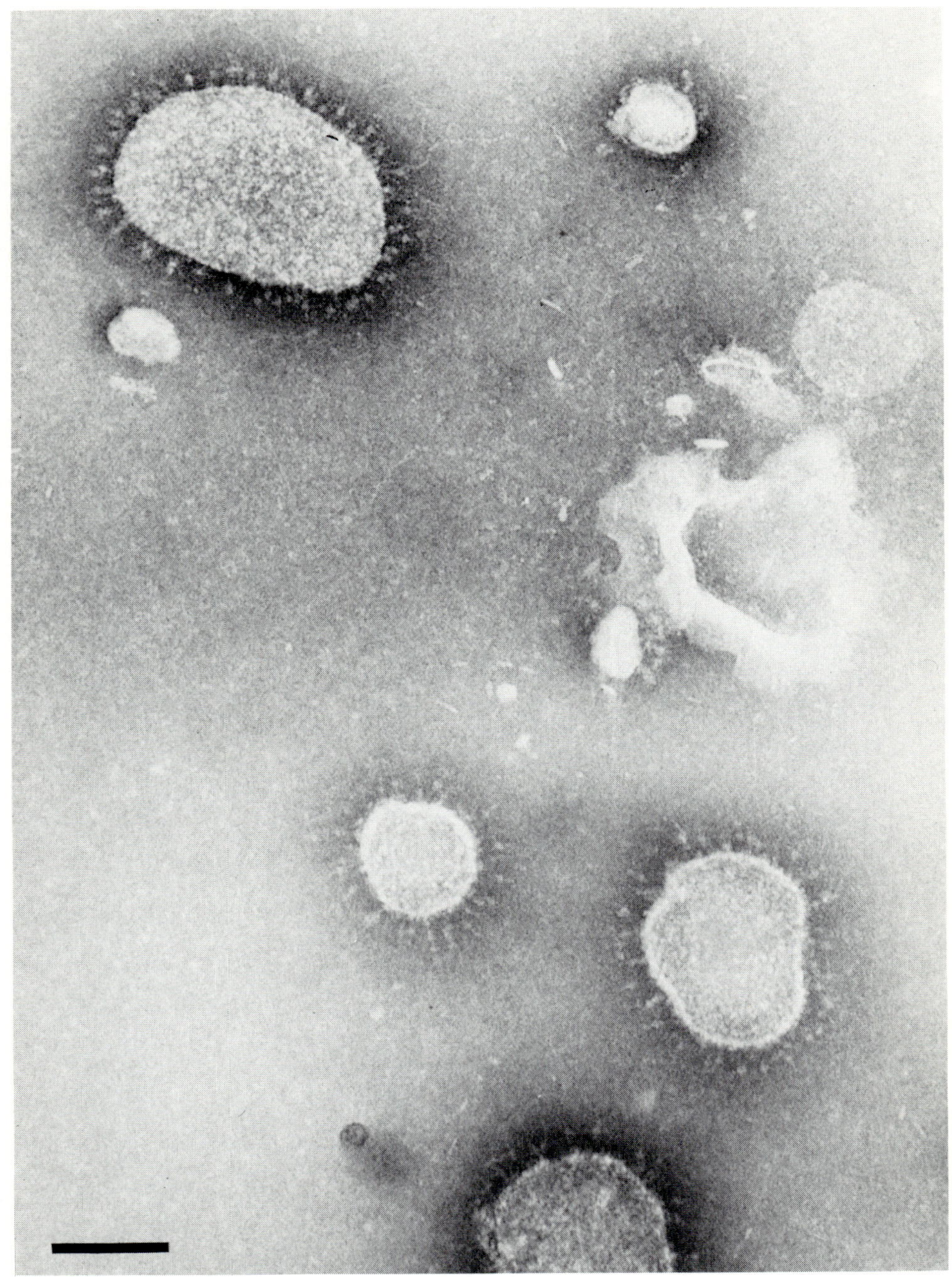

PLATE 10.   Cornavirus-like particles from stool ("fringes").

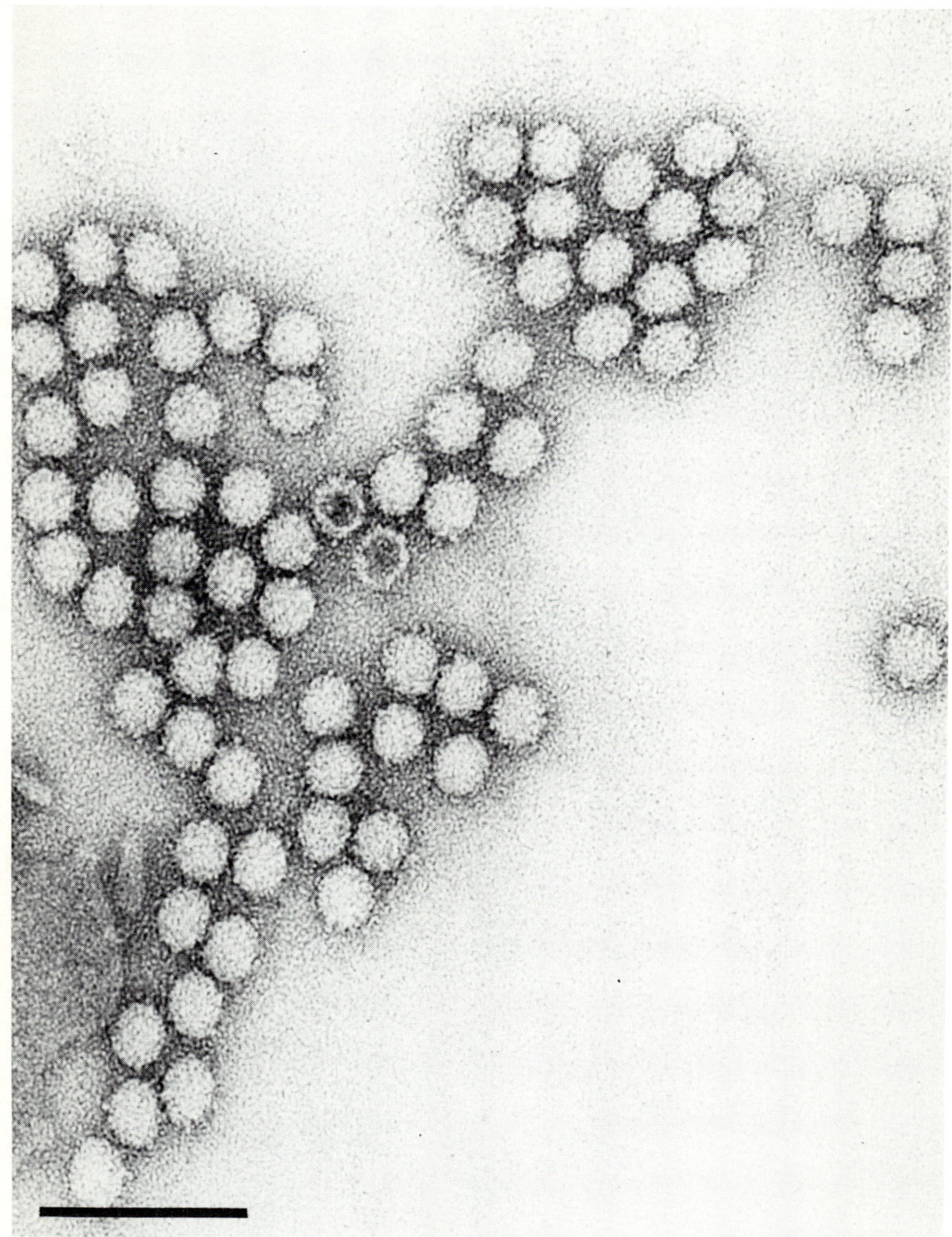

PLATE 11.   Unidentified 29-nm particles from stool.

Index

INDEX